Plumbing For Dummies

Sheet

Glossary

Use this handy glossary as a reference tool when shopping for materials, supplies, and tools.

Cap: Used to end a line of copper pipe.

Chemical welding: Process that joins rigid plastic pipes together.

Drain pipes: Collects water from sinks, showers, tubs, and appliances.

Elbow: Used to make copper pipe turn at an angle; commonly called *els*.

Escutcheon: Trim cover piece.

Main line: The passageway for the waste that comes from the toilet and from all of the sink and tub traps; also called the *sewer line*.

Main shutoff valve: Allows full flow of water through the pipe when it's open; shuts off all water to the house when it's closed.

Plunger: Tool with wooden broomstick-like handle attached to a cup-shaped piece of rubber; also called a plumber's helper. Not to be confused with a toilet plunger.

Powered closet auger: A snake that, instead of cranking by hand, has a drill-like driver that powers the shaft.

Riser: Vertical water-supply tubes.

Rough-in dimension: Distance that a toilet or sink is located from the wall.

Shutoff valve: Allows you to turn water on and off to a particular fixture or part of your house.

Slip coupling: Fitting that looks like standard couplings but slides completely over the end of the pipe.

Snake: A coiled spiral snake that's usually about 1/4-inch thick with a handle on one end; also called a *drain auger*.

Soil pipe: Large diameter pipe leading to the sewer or septic system; also called a *soil stack*.

Solder: Used for sealing joints when working with copper pipe. The pipe and fitting are heated with a propane torch until they're hot enough to melt the solder in a process called *soldering*.

Stop valve: Controls the water flow to a fixture.

Straight connector or coupling: Continues a straight run of consecutive copper pipes.

Strainer: Cup-like metal basket or disk that fits into a drain opening.

Tailpiece: The first pipe section connected to the bottom of the of the sink.

Tee: Allows you to run another copper line off an existing line.

Toilet auger: A short, hollow clean-out rod with a spring coil snake inside that has a hooked end; also called a *closet auger*.

Toilet plunger: A ball or cup-type plunger that's designed specifically for unclogging toilets.

Trap: U-shaped pipe located under the sink.

Union: Allows you to thread the new piece of pipe into the existing fittings.

Vent pipes: Removes or exhausts sewer gases and allows air to enter the system so the waste water flows freely.

Waste drain: Takes waste water out of the house.

Waste pipes: Removes water and material from the toilet.

Plumbing For Dummies®

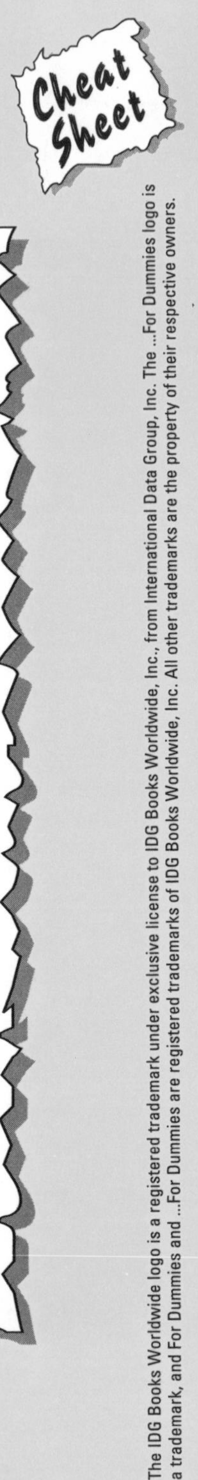

Calling a Plumber

For some tasks, you may need to call a professional plumber.

Tasks for a professional

The following tasks are best left to professionals (see Chapter 1):

- Low water pressure throughout the house
- No hot water
- Sewer line stoppage
- Frozen pipes
- Extensive water line damage

A plumbing contract

Be sure to have a clear, written agreement with your plumber before any work begins. A contract with a plumber to repair, replace, or install a fixture should include the following details:

- Description of the work to be completed
- Detailed list of materials (brand name, style, color of fixtures, or other specifications of the exact material) to be used
- Cost of material and a listing of all warranties that the manufacturer provides for the fixtures
- Cost of labor
- Job installation date

Cheat Sheet $2.95 value. Item 5174-4.
For more information about IDG Books, call 1-800-762-2974.

Praise for Gene and Katie Hamilton

"The most comprehensive guide available for those of us who are 'home improvement challenged'. . . .If you're thinking of starting a home improvement project, don't start until you've read Katie and Gene Hamilton's *Home Improvement For Dummies*."

— Susan Powell, Host of Discovery Channel's *Home Matters*

"Whether you're looking for tips on rejuvenating wood floors or culling creative ideas for holiday decorating, pay a visit to Katie and Gene Hamilton."

— *USA Today*

"Think of HouseNet as your home improvement toolbox, and caretakers Gene and Katie Hamilton as your helpful neighbors. Within these (online) pages you'll find remodeling tips, home repair tutorials, money-saving ideas, and a do-it-yourself message board."

— *Chicago Tribune*

"For home renovators looking for tips, building specs, or simply someone to commiserate with at any time of the day or night, HouseNet is the place to go."

— *Newsweek*

"Voted the best place on the Web for fixing up your home."

— Editors of Kiplinger's *Personal Finance* magazine

"The latest in home improvement information and services is just a few mouse clicks away with the relaunch of HouseNet on the World Wide Web."

— *Chicago Sun-Times*

"Before you embark on a home improvement project, check out HouseNet's cost guides to determine whether it's really worth doing yourself. The how-to section is particularly rich in detail and covers broad range of topics, from how to install an air conditioner to instructions for the first-time wallpaper hanger."

— Yahoo! *Internet Life,* New York

"HouseNet, with its magazine-style look, is a very inviting and easy-to-use home-related site. It will guide your repairs down to the smallest detail, including a calculator to help accurately figure measurements."

— *Today's Homeowner*

Praise for Carpentry For Dummies

"Don't despair over household repairs. Gene and Katie Hamilton's easy-to-read how-to manual will make you feel like you can make repairs and maintain your home like a pro!"

— Chris & Beverly DeJulio, Hosts of *HandyMa'am* on PBS and *HomeWise* on HGTV; Authors of *HandyMa'am*

"*Carpentry For Dummies* is probably the most useful tool to have in the home workshop — solid information about everything from buying lumber to finish carpentry is right at your fingertips.

— Don Geary, Author, Freelance How-To Writer, and President of the National Association of Home & Workshop Writer's

"When remodeling, it's easy to become overwhelmed by all the choices and possibilities. *Carpentry For Dummies* is a terrific guide to stay on the right track."

— Richard Karn, "Al Borland" *Home Improvement*

Praise for Painting and Wallpapering For Dummies

"The Hamiltons do everything but roll on the paint or paste the paper for you. Here is specific, detailed, and accurate information, problem-solving, and trouble-shooting. Even a first-timer will feel confident about picking up a brush and transforming a room."

— Judy Stark, Home Editor, *St. Petersburg Times,* President of the National Association of Real Estate Editors

"The thickness of a coat (only about 2/1000ths of an inch) is all that stands between your most costly investment and the environment. Before you expose your house to the next season, get seasoned advice. Brush-up with Gene and Katie Hamilton's *Painting and Wallpapering For Dummies.* It's a smart way to start your next project!"

— Thomas A. Kraeutler, Home Improvement Columnist and Radio/TV Host

"Gene and Katie have done it again. Their new book has everything. All you need now are the brushes."

— Glenn Haege, America's Master Handyman; Host of Westwood One's *Ask the Handyman* radio show

TM

References for the Rest of Us!™

BESTSELLING BOOK SERIES

Do you find that traditional reference books are overloaded with technical details and advice you'll never use? Do you postpone important life decisions because you just don't want to deal with them? Then our *...For Dummies*® business and general reference book series is for you.

...For Dummies business and general reference books are written for those frustrated and hard-working souls who know they aren't dumb, but find that the myriad of personal and business issues and the accompanying horror stories make them feel helpless. *...For Dummies* books use a lighthearted approach, a down-to-earth style, and even cartoons and humorous icons to dispel fears and build confidence. Lighthearted but not lightweight, these books are perfect survival guides to solve your everyday personal and business problems.

"More than a publishing phenomenon, 'Dummies' is a sign of the times."

— The New York Times

"A world of detailed and authoritative information is packed into them..."

— U.S. News and World Report

"...you won't go wrong buying them."

— Walter Mossberg, Wall Street Journal, on IDG Books' ...For Dummies books

Already, millions of satisfied readers agree. They have made *...For Dummies* the #1 introductory level computer book series and a best-selling business book series. They have written asking for more. So, if you're looking for the best and easiest way to learn about business and other general reference topics, look to *...For Dummies* to give you a helping hand.

1/99

by Gene Hamilton and Katie Hamilton

IDG Books Worldwide, Inc.
An International Data Group Company

Foster City, CA ◆ Chicago, IL ◆ Indianapolis, IN ◆ New York, NY

Plumbing For Dummies®

Published by
IDG Books Worldwide, Inc.
An International Data Group Company
919 E. Hillsdale Blvd.
Suite 400
Foster City, CA 94404
www.idgbooks.com (IDG Books Worldwide Web site)
www.dummies.com (Dummies Press Web site)

Library of Congress Catalog Card No.: 99-63191

ISBN: 0-7645-5174-4

Printed in the United States of America

10 9 8 7 6 5 4 3 2 1

1B/SZ/QW/ZZ/1N

Distributed in the United States by IDG Books Worldwide, Inc.

Distributed by CDG Books Canada Inc. for Canada; by Transworld Publishers Limited in the United Kingdom; by IDG Norge Books for Norway; by IDG Sweden Books for Sweden; by IDG Books Australia Publishing Corporation Pty. Ltd. for Australia and New Zealand; by TransQuest Publishers Pte Ltd. for Singapore, Malaysia, Thailand, Indonesia, and Hong Kong; by Gotop Information Inc. for Taiwan; by ICG Muse, Inc. for Japan; by Norma Comunicaciones S.A. for Colombia; by Intersoft for South Africa; by Eyrolles for France; by International Thomson Publishing for Germany, Austria and Switzerland; by Distribuidora Cuspide for Argentina; by Livraria Cultura for Brazil; by Ediciones ZETA S.C.R. Ltda. for Peru; by WS Computer Publishing Corporation, Inc., for the Philippines; by Contemporanea de Ediciones for Venezuela; by Express Computer Distributors for the Caribbean and West Indies; by Micronesia Media Distributor, Inc. for Micronesia; by Grupo Editorial Norma S.A. for Guatemala; by Chips Computadoras S.A. de C.V. for Mexico; by Editorial Norma de Panama S.A. for Panama; by American Bookshops for Finland. Authorized Sales Agent: Anthony Rudkin Associates for the Middle East and North Africa.

For general information on IDG Books Worldwide's books in the U.S., please call our Consumer Customer Service department at 800-762-2974. For reseller information, including discounts and premium sales, please call our Reseller Customer Service department at 800-434-3422.

For information on where to purchase IDG Books Worldwide's books outside the U.S., please contact our International Sales department at 317-596-5530 or fax 317-596-5692.

For consumer information on foreign language translations, please contact our Customer Service department at 1-800-434-3422, fax 317-596-5692, or e-mail rights@idgbooks.com.

For information on licensing foreign or domestic rights, please phone +1-650-655-3109.

For sales inquiries and special prices for bulk quantities, please contact our Sales department at 650-655-3200 or write to the address above.

For information on using IDG Books Worldwide's books in the classroom or for ordering examination copies, please contact our Educational Sales department at 800-434-2086 or fax 317-596-5499.

For press review copies, author interviews, or other publicity information, please contact our Public Relations department at 650-655-3000 or fax 650-655-3299.

For authorization to photocopy items for corporate, personal, or educational use, please contact Copyright Clearance Center, 222 Rosewood Drive, Danvers, MA 01923, or fax 978-750-4470.

About the Authors

Gene and Katie Hamilton are the husband-and-wife authors of the popular weekly syndicated newspaper and online column for the Los Angeles Times Syndicate, "Do It Yourself ... or Not?" The founders of www.housenet.com, the largest home and garden web site for homeowners and do-it-yourselfers, they are also the authors of more than a dozen best selling books, including *Home Improvement For Dummies.* The Hamiltons have been featured in numerous publications and have appeared on television shows, including HGTV, CNN, *Today,* and *Dateline.* In their 20 years as home repair experts, they've successfully renovated 14 homes (and they're still married!).

ABOUT IDG BOOKS WORLDWIDE

Welcome to the world of IDG Books Worldwide.

IDG Books Worldwide, Inc., is a subsidiary of International Data Group, the world's largest publisher of computer-related information and the leading global provider of information services on information technology. IDG was founded more than 30 years ago by Patrick J. McGovern and now employs more than 9,000 people worldwide. IDG publishes more than 290 computer publications in over 75 countries. More than 90 million people read one or more IDG publications each month.

Launched in 1990, IDG Books Worldwide is today the #1 publisher of best-selling computer books in the United States. We are proud to have received eight awards from the Computer Press Association in recognition of editorial excellence and three from Computer Currents' First Annual Readers' Choice Awards. Our best-selling *...For Dummies®* series has more than 50 million copies in print with translations in 31 languages. IDG Books Worldwide, through a joint venture with IDG's Hi-Tech Beijing, became the first U.S. publisher to publish a computer book in the People's Republic of China. In record time, IDG Books Worldwide has become the first choice for millions of readers around the world who want to learn how to better manage their businesses.

Our mission is simple: Every one of our books is designed to bring extra value and skill-building instructions to the reader. Our books are written by experts who understand and care about our readers. The knowledge base of our editorial staff comes from years of experience in publishing, education, and journalism — experience we use to produce books to carry us into the new millennium. In short, we care about books, so we attract the best people. We devote special attention to details such as audience, interior design, use of icons, and illustrations. And because we use an efficient process of authoring, editing, and desktop publishing our books electronically, we can spend more time ensuring superior content and less time on the technicalities of making books.

You can count on our commitment to deliver high-quality books at competitive prices on topics you want to read about. At IDG Books Worldwide, we continue in the IDG tradition of delivering quality for more than 30 years. You'll find no better book on a subject than one from IDG Books Worldwide.

John Kilcullen
Chairman and CEO
IDG Books Worldwide, Inc.

Steven Berkowitz
President and Publisher
IDG Books Worldwide, Inc.

Eighth Annual Computer Press Awards 1992

Ninth Annual Computer Press Awards 1993

Tenth Annual Computer Press Awards 1994

Eleventh Annual Computer Press Awards 1995

IDG is the world's leading IT media, research and exposition company. Founded in 1964, IDG had 1997 revenues of $2.05 billion and has more than 9,000 employees worldwide. IDG offers the widest range of media options that reach IT buyers in 75 countries representing 95% of worldwide IT spending. IDG's diverse product and services portfolio spans six key areas including print publishing, online publishing, expositions and conferences, market research, education and training, and global marketing services. More than 90 million people read one or more of IDG's 290 magazines and newspapers, including IDG's leading global brands — Computerworld, PC World, Network World, Macworld and the Channel World family of publications. IDG Books Worldwide is one of the fastest-growing computer book publishers in the world, with more than 700 titles in 36 languages. The "...For Dummies®" series alone has more than 50 million copies in print. IDG offers online users the largest network of technology-specific Web sites around the world through IDG.net (http://www.idg.net), which comprises more than 225 targeted Web sites in 55 countries worldwide. International Data Corporation (IDC) is the world's largest provider of information technology data, analysis and consulting, with research centers in over 41 countries and more than 400 research analysts worldwide. IDG World Expo is a leading producer of more than 168 globally branded conferences and expositions in 35 countries including E3 (Electronic Entertainment Expo), Macworld Expo, ComNet, Windows World Expo, ICE (Internet Commerce Expo), Agenda, DEMO, and Spotlight. IDG's training subsidiary, ExecuTrain, is the world's largest computer training company, with more than 230 locations worldwide and 785 training courses. IDG Marketing Services helps industry-leading IT companies build international brand recognition by developing global integrated marketing programs via IDG's print, online and exposition products worldwide. Further information about the company can be found at www.idg.com. 1/24/99

Dedication

We'd like to dedicate this book to all of the plumbers, professional tradespeople, and contractors who have helped us over the years. We began working on our first house short of money and skills, but long on enthusiasm. We asked questions about repairing and remodeling our little brick duplex (back in 1966), and all of the construction experts we met gave us confidence with their encouraging words. They taught us the value of using good tools, the importance of mastering skills, and the reason to practice patience. Most of all, they taught us the soothing power of laughter when all else fails. We're indebted to them all.

Authors' Acknowledgments

We've had a lot of help from professionals writing this book. Veteran home improvement writers Don Geary and Bob Gould contributed their expertise and bits of wit and wisdom. Technical reviewer Richard Perreault, a Master Plumber for 25 years, has gone over every word with a fine eye for detail to make this book as accurate and complete as it can be. We're grateful for all of their contributions.

At IDG Books, acquistions editor Holly McGuire has been the guiding light to keep us on track and even when obstacles came our way, she found a solution. On a day-to-day basis, our project editor Tere Drenth took the helm, set the course, and led us through the process, always with a smile and sense of humor.

And we'd like to acknowledge the illustrators at Precision Graphics who interpreted our words into pictures, and Lori Williams at Genova Products who shared some of their artwork with us.

And last, but certainly not least, we appreciate the help of Jane Jordan Browne, our good friend and agent, who way back in the early 1980s encouraged us to write our first book.

Publisher's Acknowledgments

We're proud of this book; please register your comments through our IDG Books Worldwide Online Registration Form located at `http://my2cents.dummies.com`.

Some of the people who helped bring this book to market include the following:

Acquisitions and Editorial

Project Editor: Tere Drenth

Acquisitions Editor: Holly McGuire

Technical Reviewer: Richard Perreault

Editorial Coordinator: Maureen Kelly

Editorial Assistant: Alison Walthall

Special Help

Kristin Cocks, Mary C. Corder, Jennifer Ehrlich, Elizabeth Kuball, Suzanne Thomas, Kevin Thornton, Michelle Vukas

Production

Project Coordinator: E. Shawn Aylsworth

Layout and Graphics: Thomas Emrick, Angela F. Hunckler, Barry Offringa, Brent Savage, Janet Seib, Brian Torwelle, Dan Whetstine

Special Art: Precision Graphics, George Retseck, Tony Davis

Proofreaders: Christine Berman, Joanne Keaton, Nancy Price, Marianne Santy, Janet M. Withers

Indexer: David Heiret

General and Administrative

IDG Books Worldwide, Inc.: John Kilcullen, CEO; Steven Berkowitz, President and Publisher

IDG Books Technology Publishing Group: Richard Swadley, Senior Vice President and Publisher; Walter Bruce III, Vice President and Associate Publisher; Steven Sayre, Associate Publisher; Joseph Wikert, Associate Publisher; Mary Bednarek, Branded Product Development Director; Mary Corder, Editorial Director

IDG Books Consumer Publishing Group: Roland Elgey, Senior Vice President and Publisher; Kathleen A. Welton, Vice President and Publisher; Kevin Thornton, Acquisitions Manager; Kristin Cocks, Editorial Director

IDG Books Internet Publishing Group: Brenda McLaughlin, Senior Vice President and Publisher; Diane Steele, Vice President and Associate Publisher; Sofia Marchant, Online Marketing Manager

IDG Books Production for Dummies Press: Michael R. Britton, Vice President of Production; Debbie Stailey, Associate Director of Production; Cindy L. Phipps, Manager of Project Coordination, Production Proofreading, and Indexing; Shelley Lea, Supervisor of Graphics and Design; Debbie J. Gates, Production Systems Specialist; Robert Springer, Supervisor of Proofreading; Laura Carpenter, Production Control Manager; Tony Augsburger, Supervisor of Reprints and Bluelines

◆

The publisher would like to give special thanks to Patrick J. McGovern, without whom this book would not have been possible.

◆

Contents at a Glance

Cartoons at a Glance
By Rich Tennant
The 5th Wave
By Rich Tennant
©RICHTENNANT
Okay, this is a little more complicated than I thought. Bring me my canoe paddle, a large balloon, and a can of spray cheese.
page 7
The 5th Wave
By Rich Tennant
©RICHTENNANT
This house came with a rather unusual septic system.
page 221
The 5th Wave
By Rich Tennant
©RICHTENNANT
OLD FAITHFUL
"Be patient everyone. This will just take a minute."
page 65
The 5th Wave
By Rich Tennant
©RICHTENNANT
Oh yeah, I see it drippin'. And if you think you're man enough to change the O-ring, why, you just go ahead. But you'd BETTER KNOW WHAT YOU'RE DOIN', BOY!
page 91
The 5th Wave
By Rich Tennant
©RICHTENNANT
"Sometimes Bill working for the city comes in real handy. Like when we decided to replace the kitchen fixtures."
page 143
Fax: 978-546-7747 • E-mail: the5wave@tiac.net

Table of Contents

Introduction

What comes to mind when you hear the word "plumbing"? If you're like most people, a chill goes down your spine as you picture water gushing out of broken pipes and toilets overflowing. The thought of actually fixing those problems yourself — and preventing them from happening in the first place — is probably not one you entertain often, and especially not when everything's working well. Plumbing doesn't usually become a priority until you're standing in six inches of murky water wondering what to do next, and then most people's first thought is, "Where can I find a good plumber?"

That's where *Plumbing For Dummies* comes to the rescue. The book you hold in your hands has the answers to every plumbing question you've ever contemplated, as well as the answers to the questions you didn't even know you had. In this book, we take the seemingly complicated and convoluted world of plumbing and put it in plain English. We make sense of the mess of pipes running into, out of, and throughout your house. And we give you the power to tackle plumbing problems yourself — as well as the insight to know when you need a plumber.

You use water in your house in many ways that you probably never even think about, and knowing how it all works can save you a bundle, because plumbing repairs and improvements can be costly. Not having to depend on a plumber and being able to fix a leaky pipe or dripping faucet is a real benefit. Even if you don't want to get plumbing putty on your hands, flipping through these pages you can pick up some useful plumbing know-how that'll help you sound knowledgeable when you have to call a plumber.

Plumbing For Dummies contains all of the information you need to know to get started and much more — including not only plumbing basics but also the detailed know-how required to make fairly major repairs and replacements. From the mundane chores of unclogging a toilet or drain to the more satisfying jobs of replacing a kitchen sink or bathroom vanity, you'll find everything you need to be a master of plumbing projects throughout your house. *Plumbing For Dummies* can help you transform yourself from someone who only looks under your bathroom sink to reach for a new roll of toilet paper to someone who understands what all those pipes do and how to fix them when a problem arises.

Who Needs to Read This Book?

Do you hate nothing more than having to call (and pay) a plumber when your bathroom sink is draining too slowly or when your toilet won't flush? Do you like the thought of tackling a project like installing a new sink and giving your kitchen or bathroom a brand-new look? Do you long to know the names of the various valves and pipes that make up the plumbing system in your home? If you answered yes (or even maybe) to any of these questions, then *Plumbing For Dummies* is just the book for you.

To understand the information in this book, you don't need any knowledge of plumbing whatsoever. (All you need to know is that water doesn't flow out of your faucets by magic.) We don't assume you have any plumbing know-how beyond buying a bottle of liquid drain opener and pouring it down your drain. But if you've done some fiddling around with minor plumbing problems already, you'll still find lots of great tips and techniques here. So whether you're a complete novice or you've already gotten your feet wet in plumbing projects in the past, *Plumbing For Dummies* is the book for you.

How to Use This Book (Other Than as a Doorstop)

You can use *Plumbing For Dummies* in many ways, including the following:

- **If you want information about a specific topic, such as how to cut plastic plumbing pipe, you can flip to the Index and find the relevant page or pages that cover that topic.**
- **If you want general information about a topic or project, such as how to unclog a shower drain, just flip to that chapter and read what you need.** The chapter headings and icons in the margins direct you to the information you want. Occasionally, we refer you to related information in other chapters, too.
- **If you want to cover all your bases, read the whole book.** You'll discover options that you never knew you had. Although you certainly don't need to keep all this information in your head, you'll be surprised at what you discover. And you can dust the book off and consult it as you tackle your next project.

How This Book Is Organized

Plumbing For Dummies is organized in five parts, and the chapters within each part are broken down into sections, which cover specific topics in detail. We begin the book with the basics of how a plumbing system works and give you a solid foundation in the tools and techniques used in plumbing work. Throughout the rest of the book, we give you the nitty-gritty details of completing the most popular and common plumbing repair and replacement projects.

Part I: Becoming Your Own Plumber

Just the facts, ma'am, that's what you'll get here — the basics of how the plumbing system made up of pipes, drains, and valves throughout your house works. Even though you can't see the maze of pipes and drainlines behind the walls, we help you understand how and why they do what they do.

If the endless aisles of plumbing fittings, sealants, gadgets, and supplies at the local home center overwhelm you, we help demystify the shopping experience so you can walk confidently into the store, with your head held high, and find exactly what you need.

The old carpentry adage "Measure twice, cut once" applies to plumbing, too. In this part, we also walk you through the basics of plumbing tools and how to use them. With this know-how, you'll be able to tackle plumbing repair and replacement projects with confidence throughout your home.

Part II: Opening Clogged Drains with Friendly Persuasion

All the chapters in this part help you through some of life's little plumbing traumas. Any clogged drain is worse than a bad hair day (and sometimes that hair is the cause of the clogged drain!), so we show you and tell you how to fix it. Yours won't be the first (or last) dirty diaper to clog a toilet, but when it happens at your house, it sure isn't fun. The chapters in this part give you the answers to these and many more compelling, everyday plumbing mishaps. So even though you're sure to encounter some of these problems in the future, you can laugh them off (or at least get through them with a smile), knowing that you have the knowledge and power to fix them.

Part III: Repairing Leaks Like a Pro

If it leaks, fix it! The chapters in this part run the gamut from quick fixes to long-term repairs for pipes, faucets, and toilets, so you can do battle with any leak life throws your way. We cover all kinds of leaky pipes — everything from a simple leak to a more challenging one from a frozen pipe. We also take you step-by-step through repairing all kinds of faucets, bathtubs and showerheads, and the ever-popular toilet.

Even if you don't want to do the repair work yourself, the information in these chapters is important, because when you call a plumber, you'll not only be able to make sense of the technical terms your plumber uses, you'll also know how to diagnose the problem yourself (so you're sure you're not getting charged for repairs that you don't need).

Part IV: Tackling Plumbing Projects

Often, all you need to do to transform a bathroom is get rid of an old, battered vanity and install a pedestal sink. Or maybe a new, stainless, double-compartment sink and faucet is the winning ticket to make your kitchen shine. Plumbing projects like these are the most popular of all, because they have the power to make your home look better and work better. So these chapters feature step-by-step instructions for replacing the basic fixtures in kitchens and bathrooms. We lead you through some great projects from start to finish, giving you all of the information you need to tackle them on your own.

Part V: The Part of Tens

No *...For Dummies* book would be complete without The Part of Tens. And we think that you'll find the chapters in this part especially useful. One chapter highlights especially valuable information, including tips for maintaining a sewer system. Another chapter provides new and important information on saving water, while a third points you to some hot plumbing Web sites.

Icons Used in This Book

This wouldn't be a *...For Dummies* book without those friendly little icons in the margins. Icons tell you something about the words that they sit next to, so take a look at the following to get a grip on what the icons look like and what kind of information they point to.

Get on target with these great time-saving, money-saving, or sanity-saving inspirations from people who have been there and done that.

We don't bombard you with a bunch of technical trivia in this book, but we think some of these tidbits are too good to pass up. Some of the information is helpful; some is just interesting (at least to us). Understanding the *why* often eliminates the *what, where, when,* and *how* and clears up other questions that would otherwise fill your brain. But like so much of the data in this Information Age, you can live without it.

This icon points out key pieces of plumbing information that you need to remember.

Plumbing occasionally involves poisonous chemicals, and some of the techniques we describe can be dangerous. But you can rest assured that we alert you to these potential hazards with the Warning icon.

Some products on the market do such a darn good job that we can't resist sharing them with you. This icon points you to these products. We also use this icon to share some cool gadgets and gizmos that can more your task easier (or your toolbox more cluttered). Rarely are these tools absolutely necessary, but who cares?

Whenever we think that a plumbing repair or installation is too complicated or not within the realm of a do-it-yourselfer, you see this icon. In many situations, even a handy homeowner is best advised to hire a plumber who has the skills, tools, and experience. Sometimes, building codes actually *require* a licensed plumber do the job.

Where to Go from Here

What do you do now that you have *Plumbing For Dummies* at your side? Go to work! Try some of the techniques you read about, buy some new gadgets and gizmos, or eagerly await the next leaky faucet. With the plumbing know-how in this book, you're ready to tackle any problem or project that comes your way.

Part I

Becoming Your Own Plumber

In this part . . .

In this first section, you find out about the system of pipes, drains, and valves that run throughout your house. We help you understand how nice, hot water squirts out of the shower on demand, where the toilet water goes when it's flushed away, and many more mind-boggling mysteries.

We also make suggestions about the kinds of projects you may want to tackle and those that are best left to a professional. For those that you decide to tackle yourself, we give you a primer on building codes and permits.

Finally, we know that it can be overwhelming and downright scary not knowing what kind of sealant to buy or how on earth to use solder. So, we help you walk confidently into the plumbing department of even the largest home center and not be intimidated by the materials, supplies, and tools.

Chapter 1

The Plumbing Do-It-Yourselfer

In This Chapter

- Doing plumbing tasks yourself
- Hiring a plumber when you need to
- Understanding building codes

Considering the range of plumbing-related replacements and repairs, a do-it-yourselfer can find plenty to do — and that's what you see throughout this book, plumbing projects that are doable and in most instances, repeatable. Plumbing skills are sort of like computer skills. After you know how to format a proposal or generate a spreadsheet, you can do it again and again, although it may have seemed difficult at first. Well, after you unclog the kitchen drain, you can use that skill (and the tools that you acquire in the process) many more times. You can then confidently tackle a stopped-up bathroom drain and before long, become the neighborhood know-it-all for solving plumbing problems — all because you took the first step of plowing through the shreds of avocado in your kitchen sink disposer.

Figuring Out When to Do it Yourself

There are probably few plumbing jobs that you can't figure out how to do with a little know-how, a knowledge of the rules (codes), and some spare time. Some of the easiest plumbing jobs — mostly preventive maintenance — to do yourself are as follows:

- **Clearing slow-draining sinks and tubs:** This usually involves using a plunger or opening a trap to remove hair, food, or paper. You may need to use a hand auger to get the proper flow of water through the drain. Tub drains are harder to clean than sink drains, and it may take several tries to get the pop-up drain stopper properly adjusted, but you can do it yourself. (The chapters in Part II cover a variety of ways to unclog drains.)

- **Repairing or replacing leaky valves:** The steady drip-drip-drip of a faucet is certainly annoying, but it's a repair that can wait until the weekend, so that you have time to get your hands on the correct repair parts before you begin work (Chapter 10 shows you how). This may even be a good time to replace your old faucet with one of those fancy new ones (see Chapter 15).

  If you don't already have shutoff valves (covered in Chapter 2) on the water supply lines under your sink, this is a good time to add them. Future repairs will be much easier when you can shut off the water to individual sinks or appliances. You won't have any questions, such as "Dad, when can I wash my hair?" that can happen when the main water shutoff valve (also discussed in Chapter 2) is closed for plumbing repairs in the house.

- **Solving toilet-tank problems:** Whether the problem is a valve that won't shut off or a tankball that won't seat properly, these are repairs that can wait until you have the time to work on them (see Chapters 7, 12, and 17). But don't wait too long. A small leak can waste a lot of water.

- **Maintaining the washing machine:** Don't forget to check the hoses. If they've been in use for several years, change them. Purchase a set of replacement hose that have a braided reinforced cover to prevent the hose from bursting. The small extra cost is worth it; if one should burst, you'll be calling for help from more pros than just the plumber.

- **Preventing winter freezing:** You can prepare for winter by insulating water and drain lines that are subject to freezing, or wrapping them with automatic heat tape. (Don't use heat tape on plastic pipe, though — eventually, the pipe will deform.) Read the directions on the package: Suggested applications for the tape and its electrical requirements are clearly described. This prep work may save you from huge messes after a sudden, subzero freeze.

Knowing When to Call a Pro

Don't lull yourself into believing that you'll never need a plumber. If you have a plumbing emergency, there are two things you and your family need to know:

- The location of the main water shutoff valve (see Chapter 2).
- The name and phone number of a reliable plumbing repair company.

In addition to using your local plumber for occasional emergencies, the following tasks are best left to professionals:

- **Low water pressure throughout the house:** Several factors can cause this problem. It may be caused by obstructions (rust or debris) in the water lines, which can start at the meter and run all the way to the faucet *aerators* (small strainers on the end of the spigot). It may be low water pressure from the city supply or a well. It may even be poor supply-line design. A good plumber knows how to analyze the problem.
- **No hot water:** It's obvious *what* happened, but unless the hot water tank is leaking, it may take awhile to find out *why.* If the tank is electric, it could be a bad heating element, a tripped circuit breaker or blown fuse, a faulty thermostat, or a bad overload switch. On gas heaters, thermocouple burners, and igniters can fail.

  No one likes to be without hot water for long. Your grandmother may have heated bath water on a stove, but that's not the way people do it today. Call your plumber for this one — he or she has had a lot of experience and can tell you if you need a new heater or if it can be repaired. If the heater needs to be replaced, your plumber will carry the new one to the basement, hook it up, make sure that it works properly, and dispose of the old one. If you decide to tackle this one yourself, check out Chapter 18.
- **Sewer line stoppage:** If you've tried all of the tricks you know to get your sewer line to drain properly, yet backups continue, you probably have a bad plug (it's often caused by tree roots) in the line that runs out to the main sewer. Rather than rent one of the big sewer rodding machines that you may break — or that may damage your sewer — call a plumber or drain cleaning service. If they get in trouble, they'll make the repairs. If you have a septic system, flip to Chapter 19 for tips on maintaining it.
- **Frozen pipes:** If a pipe freezes, before attempting to thaw the pipe, close the main water shutoff valve (see Chapter 2) and open a faucet nearby. Check carefully to see if the pipe has already burst or cracked. If it's bad news, you may need a plumber. If not, hair dryers and heat guns are the safest way to thaw a pipe. If you must use a propane torch, do so with great care — old, dry wood (which usually surrounds pipes) catches fire easily. Even if the pipe isn't burst or cracked, you may still want to call a plumber — some plumbers simply replace a section of frozen pipe rather than thawing it.
- **Extensive water line damage:** Usually caused by freezing, repairing the problem can take much of your valuable time. It's better to pay the plumber so you can earn money at your regular job.

Hiring a licensed plumber

Your local government wants to limit who can mess around with the public sewer and water lines, so they offer a *license* to plumbing contractors who have passed a rigorous test.

Find your plumber at a party

We find that the best place to locate any contractor — including a plumber — is at a neighborhood party. Why? Because word of mouth is still the best source for a referral that you can get. Who better to work on your house, than someone who has worked in your neighborhood where the houses are often the same vintage? (Read that as "similar plumbing systems.")

Here's a distinction to remember: All licensed plumbing contractors are plumbers, but not all plumbers are licensed. A plumbing contractor can have other (sometimes unlicensed) plumbers working under his or her license — the licensed plumbing contractor is responsible for the workmanship and quality of his or her workers. This license may be revoked if the plumber has a history of doing shoddy work. Most professionals take pride in their work and value their licenses.

If you consider hiring a plumber, make sure he or she is licensed. This is especially important if you're hiring an individual to do a side job for you or work on the weekends — remember that you don't have much recourse with an unlicensed plumber if the work is unsatisfactory.

Working up a plumbing contract

Just as important as finding the right plumber to do the job is having a clear, written agreement with that plumber before any work begins. The agreement doesn't have to be complicated, but should contain some basic points. A contract with a plumber to repair, replace, or install a fixture should include the following details:

- Description of the work to be completed
- Detailed list of materials (brand name, style, color of fixtures, or other specifications of the exact material) to be used
- Cost of material and a listing of all warranties that the manufacturer provides for the fixtures
- Cost of labor
- Job installation date
- Amount of deposit, if required

If your plumbing project is in an old house, hire a plumber with old-house experience, not a plumber who works strictly in new construction.

Understanding Building Codes, Permits, and Inspections

Building codes are guidelines set down to assure the safety and building standards of new and remodeled buildings. They cover all the components of a building, including its plumbing lines, by specifying the minimum standards for materials and methods used. Codes go hand-in-hand with a building permit and inspection (covered in the "Permits and inspections" section, later in this chapter) to guarantee that good workmanship was performed and quality materials were used.

The most important thing for a do-it-yourselfer to know about building and plumbing codes is that they exist, they are very important, and they must be followed when repairing, remodeling, or building a home. It's not possible for a homeowner to know all the interpretations of these codes, but building inspectors and licensed plumbers can provide the answers.

Most of the United States is governed by rules made by one or a combination of three building code organizations — ICBO, SBCCI, and BOCA. (In the future, there is supposed to be one code, the International Building Code.) Each community has the opportunity to modify these codes to suit its own special situations and idiosyncrasies. Unfortunately, though, the rules don't always seem too logical.

Don't count on the big chain home centers carrying the proper materials for the area they're located in. These big chains order supplies for all of their stores, and one store may carry materials that aren't acceptable in your community. Small hardware and plumbing supply stores are usually more knowledgeable about code issues. The final authority, though, is a city code official.

The right side of plumbing codes

We can't possibly include all the rules, so it's up to you to find out what the rules are where you live so you don't make any errors. Before you plan any plumbing job, first check with the local building codes to make sure that you know the requirements. A copy of your local building code can be purchased from you local town government: Go to the town hall and find the building department. You may find a lot of technical jargon in this document so consider having a discussion with the plumbing inspector in which you explain your project. In most cases he or she can give you the necessary advice to keep you on the right side of the code.

Permits and inspections

When do you need a permit? Well, that depends. In general, if you're replacing a plumbing fixture, faucet, or appliance, no permit is required. If you're relocating the fixture, faucet, or appliance, though, you may need a permit. If you're adding or extending new plumbing lines or installing a major appliance like a water heater, it's likely you'll need a permit.

To obtain a permit, you have to fill out a simple form that describes the work and its estimated value. If you're doing the work yourself, you indicate this; otherwise you supply the names and license numbers of the contractors that you will hire. On large jobs, the town may require that you or the contractor post a bond, which guarantees that the work meets the code (if you don't, the contractor or you may have to forfeit the bond money). In our town, a building permit and inspection goes for $35.

Your best bet is to get a copy of your local building codes (or just the plumbing requirements part of the code) from your local building department so that you have a clear-cut explanation of your requirements. Your building department is under your local government listing in the phone book.

We've had our ups and downs with building inspectors, but in most cases they're there to help, and not hinder or halt the work that you want to do.

Depending on the extent of the project, you may have to have two inspections: one after you've worked on plumbing pipes and lines (behind the wall) and a final inspection when the wall is closed in. Keep this in mind when you're scheduling your work and the inspections.

Chapter 2

The Plumbing System in Your Home

In This Chapter

- Understanding how water runs through your house
- Getting information about your water meter
- Figuring out your drain-waste-vent lines
- Understanding the importance of shutoff valves throughout your house

Before you begin fixing leaky faucets or unclogging drains, you need to know how water gets from one place to another in your house. This chapter covers the basic plumbing system of a home, introducing you to the water meter and the gate valve. You also get a grip on the drain-waste-vent (DWV) lines so you can toss that acronym around amongst your buddies and feel like a real pro.

This chapter also covers the location of the main shutoff valve and the many other valves that are a part of your home's plumbing system. These valves allow you turn off the water supply to the particular fixture or appliance that you're working on or having problems with. So if you want to replace a sink in the bathroom, for example, you don't have to shut off the water supply to your entire house while you do the repair. And your family will thank you for that. Almost all houses have a shutoff valve wherever the main water supply comes into the house, with individual shutoff valves at toilets, sinks, bathtubs and showers, and appliances like dishwashers or washing machines. If your house has outdoor spigots, you may need to locate the valves for those as well. Fear not — this chapter has all the information you need to locate the valves in your house before you make large or small repairs.

The Water Runs Through It

You can easily understand the plumbing system of your house if you keep a few basic facts straight:

- The skinny pipes bring the water in, and the fat pipes carry it out.
- Leaks in the skinny pipes, or in anything attached to a skinny pipe, can flood a house.
- Leaks in the fat pipes cause your house to stink.

Okay, plumbing is a little more involved than that, but if you think of your house as having two different water systems — one that brings fresh water in the house and another that takes waste water out to a sewer or septic system — the maze of pipes throughout your house (see Figure 2-1) may start to make some sense.

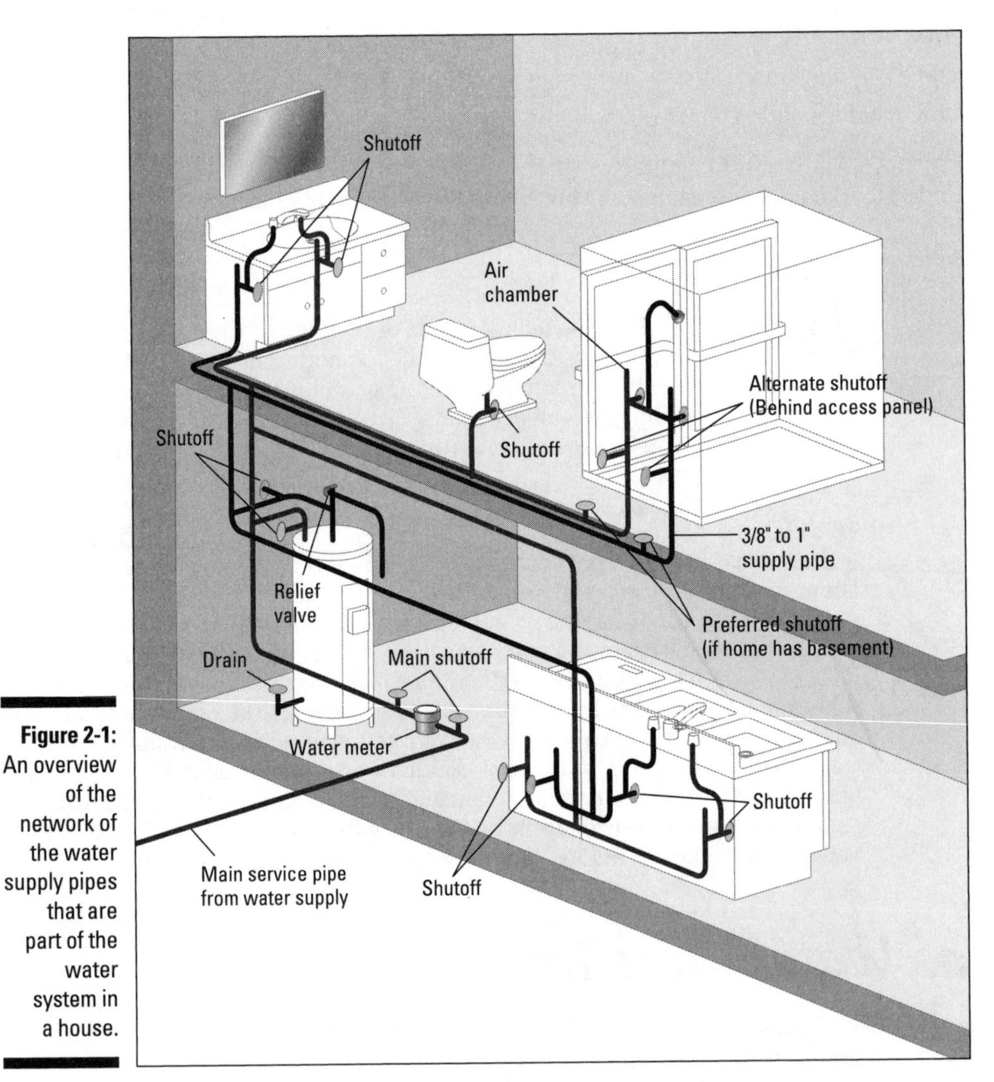

Figure 2-1: An overview of the network of the water supply pipes that are part of the water system in a house.

Plumbing really isn't complicated, and modern plumbing materials like plastic pipes and *fittings* (such as elbows and unions) allow you to successfully tackle home plumbing projects that only a decade ago would have been considered too difficult. Recognizing the growth of the do-it-yourself market, manufacturers of plumbing tools, fixtures, and materials are constantly improving the packaging and instructions for them. See Chapters 3 and 4 for a complete rundown of these products.

Most building departments now allow homeowners to do plumbing work on their own houses. Of course, regardless of who does the work, it still must meet the local building codes. So we provide important information in Chapter 1 about how to make sure that the work of you or a plumber you've hired will pass the code.

Getting Water from the City to Your House

Water running through the municipal water main is carried to your house by tapping into the water main and attaching to a smaller pipe that leads underground from the main into your house. The water typically passes through three valves and a water meter on its trip to your house — this section describes how this system works.

Three valves help bring the water to your house:

- **Corporation stop:** The valve at the municipal water main that's buried underground, usually under your street. The valve is used to turn off the water from the main to curb stop. You'd have to dig a hole in the street to get to this valve, so you need to hire a professional if your plumbing problem is in that valve.
- **Curb stop:** A valve that's somewhere between the corporation stop and your house. This valve is controlled by the municipality and is usually located close to the street. It may be in a chamber holding the water meter or in a cast iron sleeve called a *buffalo box.* In northern locations, the curb stop is buried deep in the ground to prevent freezing — you will need a special, long-handled wrench to turn this valve on or off. In more moderate climates, this valve will be located closer to the surface. Look around your property and find out where this valve is located. (The cast-iron head is a real lawnmower buster.) If you ever face a leak in the pipe between the street and your house, this is the valve to turn off. Okay, you may not be able to turn it off yourself, but you can call the city or show the plumber where it is.

- **Meter valve:** The third valve is inside your house. This valve is usually installed just before the water meter. Whoever installed the water meter installed the valve so the meter can be removed for servicing should it fail. This valve is located before the water meter so it may be outside your house if the water meter is outside.

The main shutoff valve

After the water passes through the three city-installed valves, it comes to the what we call the *main shutoff valve*. This valve is usually in the basement or on an outside wall in a utility area of the house. The main shutoff valve allows full flow of water through the pipe when it's open. Turning off this valve (by turning it clockwise) cuts off the water supply to the entire house.

The main shutoff valve may have one of two designs:

- **Gate valve:** Gate valves are very reliable and last for years, but they become difficult to turn after not being turned for years. If you haven't closed the main shutoff valve since you moved into your house, do it now. Better to find out that you can't turn it with your bare hands now, rather than waiting until you're standing in 6 inches of water.
- **Ball valve:** Houses with plastic or copper main water pipes leading into the house may have a full-flow ball valve. This valve is open when the handle is aligned with the pipe. To close it, turn the handle clockwise ¼ turn so it is at right angles to the pipe.

The main valve is the one to stop most plumbing catastrophes, like a burst pipe. Be sure everyone in the household knows where this valve is located and how to turn it off. Turning the handle clockwise closes the valve (see Figure 2-2). You need to turn the handle several turns to fully close the valve. After you've closed and opened the valve, it may start to leak a bit around the valve stem. The stem of the valve is held in place with a packing nut. Tighten this nut just enough to stop the leak. Don't overtighten it or the valve will be difficult to turn. (If you need a cheat sheet to remember which way to turn the control, use a label or tag with the simple reminder: "Right off" with an arrow pointing right, for example.)

Any time you shut off the water and allow the pipes to drain, unscrew the *aerators* (small screens on the end of all faucets) before you turn the water back on. Doing this prevents the small particles of scale that may shake loose from inside the pipes from clogging the small holes in these units (see Chapter 5 for more information on unclogging an aerator).

In the "Locating Shutoff Valves" section, later in this chapter, we cover other shutoff valves throughout a house.

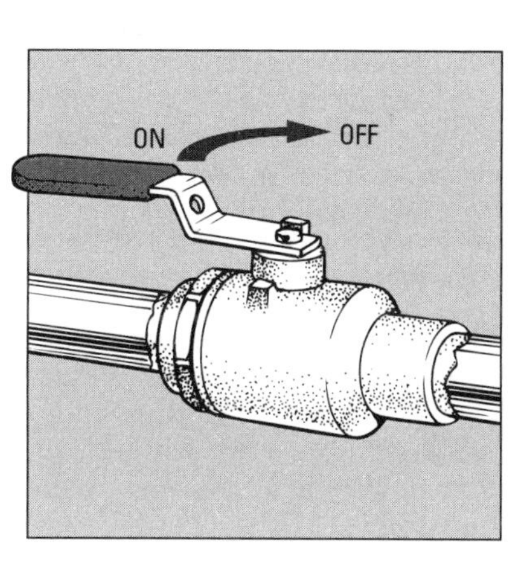

Figure 2-2: Label a tag on all shutoff valves showing an arrow pointing to the right (clockwise) to shut it off and to the left (counterclockwise) to turn it on.

The water meter

If there is any part of the water supply system that you wouldn't mind breaking, it's probably the water meter. Unfortunately, as water meters begin to fail, they usually run fast and record too much water usage, not the other way around. And besides if your water meter stops, your local water company will just prorate your water usage to some average.

The meter may be located streetside and is owned and monitored by the water company. Sometimes the meter is buried in the ground and covered with a popup lid; it's usually located near the front of the house or on the side of the house near the street. In cold climates, meters may be located inside the house, but most municipal water companies have modified inside meters so they can be read from the outside of the building. By using plain-old arithmetic, the utility company can figure out how much water your household has consumed. They compute the difference between last month's reading and this month's reading. Sometimes meter readings are done on an estimated basis; other times meter-readers prowl through your neighborhood with little handheld gizmos where they input the readings on the meters.

Reading a water meter

Most meters have a cover that opens when the hinged top is raised. The meter has a digital display or round dial that shows the number of gallons that have passed through the meter. The statement for a water bill reflects the total number or gallons used between the billing periods.

On the monthly or quarterly statement that comes from the water company or utility, you can usually find an explanation of how to read the meter so you can keep track of water usage yourself. Call your local utility company and ask for directions if you're not sure how to read your own meter.

By comparing the reading of the water meter over a very short time, you can spot a major leak. If you turn off all of the faucets (indoor and outdoor), along with any appliances that may use water (like a dishwasher or washing machine), no water should be flowing through the meter. To check for a major leak, make sure everything is off, and then record the meter reading. Check the meter in a few minutes and it should read the same. If the dials have moved, you have a leak someplace. This technique isn't a good check for a minor leak like a dripping faucet. Eventually the meter records the water used, but it is so small that it's difficult to notice any meter movement. Don't let the apparent lack of movement in the meter fool you. Over time even a small drip, drip, drip wastes hundreds of gallons of water, which is why we devote Part III of this book to fixing leaks.

Because you pay for water that passes through the meter, any leak past the water meter can cost you big bucks. If water leaks from the pipes that lead to your house, it doesn't pass through the meter and therefore can't be counted by the water company. Any leak, however, can do damage to your house even though it may not raise your water bill.

In most areas, your sewer bill is based on your water use. Some cities, however, give homeowners the option of installing two water meters: one for indoor household use for which sewer rates are charged accordingly, and the other for a sprinkler meter that feeds outdoor hoses and lawn sprinkler systems for which no sewer rate is charged. Not a bad idea, huh? Saves lots of money!

Figuring Out Your Drain-Waste-Vent (DWV) Lines

The fat pipes in your house make up the *drain-waste-vent system* (also known as the DWV) and carry wastewater to a city sewer line or your private sewer treatment facility (called a *septic tank and field*).

- The *drain pipes* collect the water from sinks, showers, tubs, and appliances.
- The *waste pipes* remove water and material from the toilet.
- The *vent pipes* remove or exhaust sewer gases and allow air to enter the system so the waste water flows freely.

Just imagine what the typical water system in your house does day in and day out!

The drain pipes are made of cast iron, galvanized pipe, copper, or plastic (see Chapter 3 for more on materials). Local building codes that regulate the materials used in the DWV system have changed over the years, so most older homes have a combination of materials.

A typical bathroom sink is a good example of how these components all work together. You've probably never spent much time observing the pipes beneath your vanity, but take a look and here's what you'll see.

- Water runs down the sink drain into a *p-trap* (called that because it's curved like the letter), which fills up with water to prevent sewer gases and odors from getting into the house through the pipe. This water gets refreshed whenever more water runs through it.
- A drain pipe attached to the p-trap goes into an opening in the wall. This is the last of what you can see.
- Behind the wall, a vent line and drain pipe lead to a soil stack, which is the control central of the water system. Drain pipes take the wasted water to the soil stack; through the stack, sewer gases are carried up to the roof through vent lines.

All the faucets and water appliances in a house use this same system of drains, pipes, and vents. All the waste lines have a cleanout, which is a Y-shaped fitting that's accessible to clean out any really serious obstructions within the system. (See Figure 2-3.)

Locating Shutoff Valves

The *shutoff valve* allows you to turn water on and off to a particular fixture or part of your house. In general, all shutoff valves are below or near the source of water. That makes perfect sense, right? But unless you know where to look, the obvious isn't always so obvious. Take a little tour of all these rooms and do some serious snooping around to find the valves in your house.

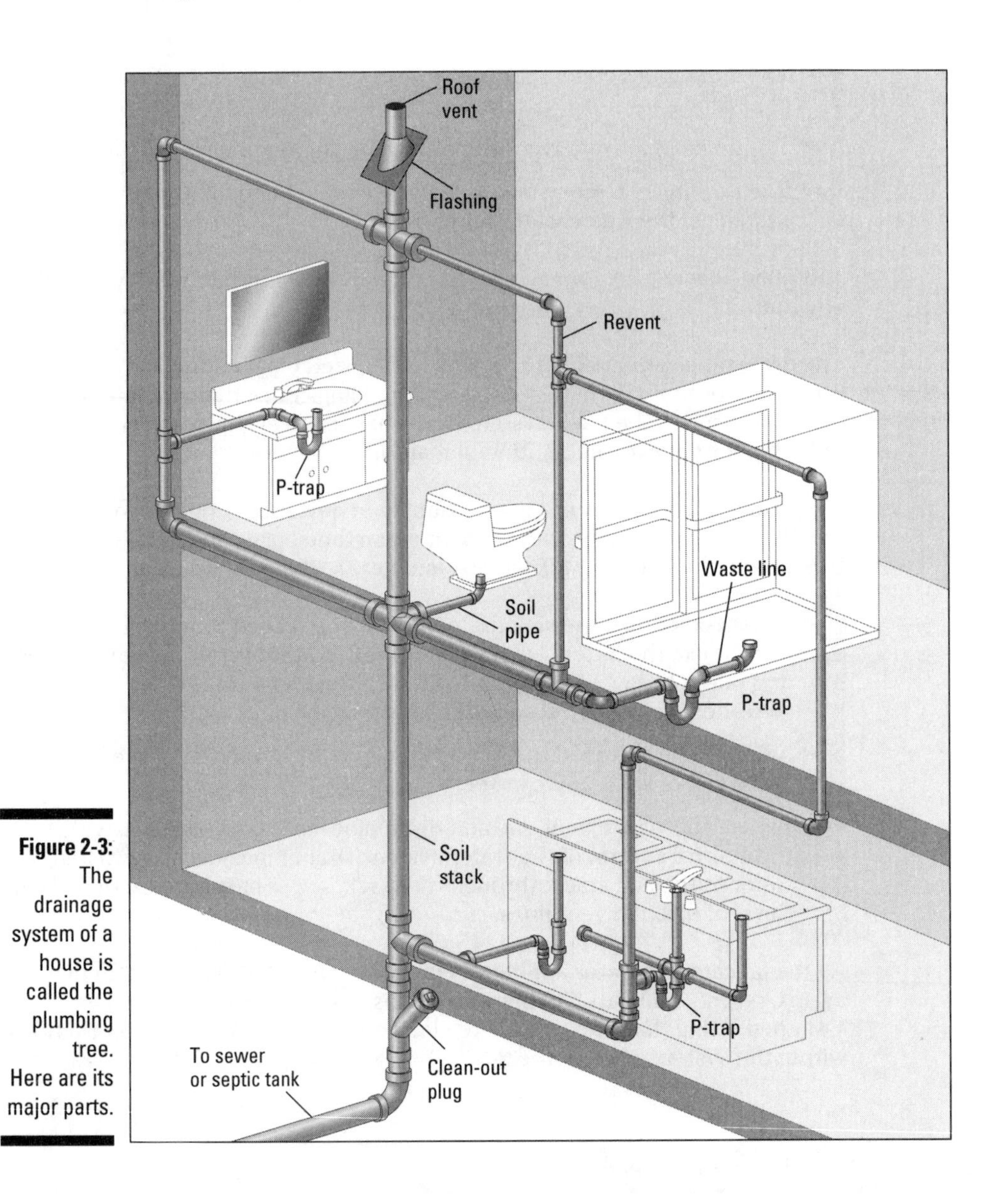

Figure 2-3: The drainage system of a house is called the plumbing tree. Here are its major parts.

In the bathroom

Most bathrooms have shutoff valves that are designed to control the water flow to the toilet, sink, and bathtub or shower. These fixture shutoff valves are also called *stop valves,* because they stop the water from getting to the fixture. (Hey, some plumbing terms make sense!) Knowing where each of these valves is located can save you time and energy when you're doing plumbing repairs. We tell you just where to find them in the following sections.

The toilet

The shutoff or stop valve beneath the toilet is one of the more accessible valves, because it comes out of the wall or floor and is clearly visible — see Figure 2-4. In fact, you've probably seen it many times without noticing it. Just look for a handle behind your toilet. And that's all there is to it!

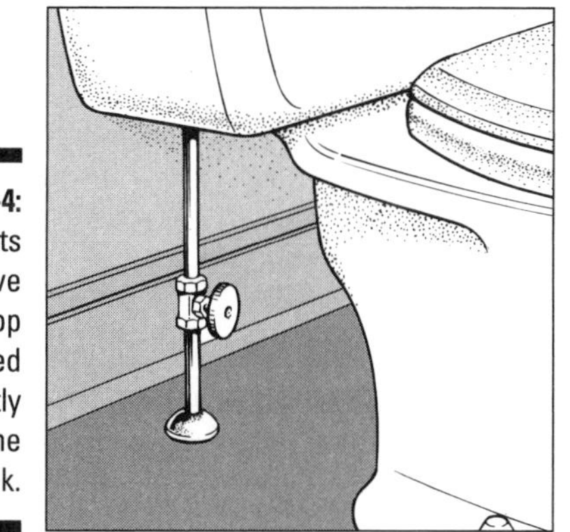

Figure 2-4: Toilets usually have a stop located directly below the tank.

Sink or vanity basin

A wall-hung or pedestal sink has shutoff valves that are clearly accessible, but sinks in a vanity cabinet aren't always easy to find or reach. We all know we should keep the cabinets under our sinks organized, but, let's face it, that's easier said than done.

So go on a cleaning binge just once, organize the bottles of shampoo and rolls of toilet paper, and find the shutoff valves. Your sink has two valves — one for the hot water and one for cold water (see Figure 2-5). If you're lucky, the valves won't be tucked up high behind a cabinet partition. But if they are, you may need to lie on your back and shine a flashlight to find them. (Plumbing: It's always an adventure!)

The bathtub and shower

Finding the shutoff valves for the bathtub or shower may take a little snooping. In most houses, one of the walls in a bathroom contains all of the plumbing pipes. So look for the bathtub and shower shutoff valves on the opposite side of the wall that contains the plumbing pipes.

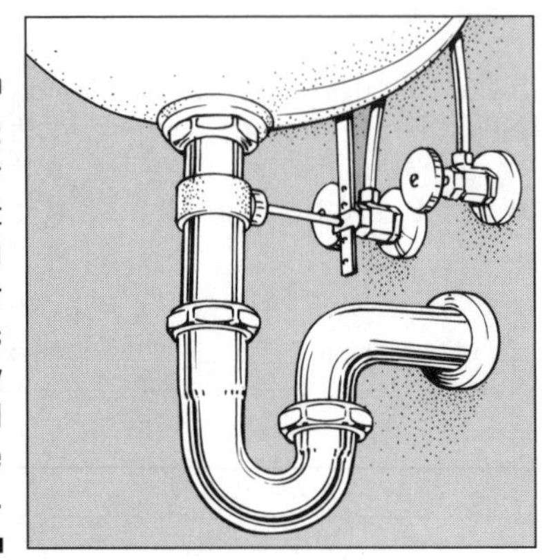

Figure 2-5: A shutoff for both the hot and cold water supply is usually located below the sink.

"How do I know which wall contains the plumbing pipes?" you may ask. The wall that contains the pipes usually has a removable panel on the opposite side, which covers a recess in the wall and hides the pipes from view. The panel may be located in a closet, a bedroom, or a hall that is adjacent to the bathroom. So just snoop around in your closets and look for a panel like the one shown in Figure 2-6. After you find it, remove the screws holding the panel in place, and you see the backside of the tub or shower. You may or may not find a set of shutoff valves in this wall recess. (Frustrating, isn't it?)

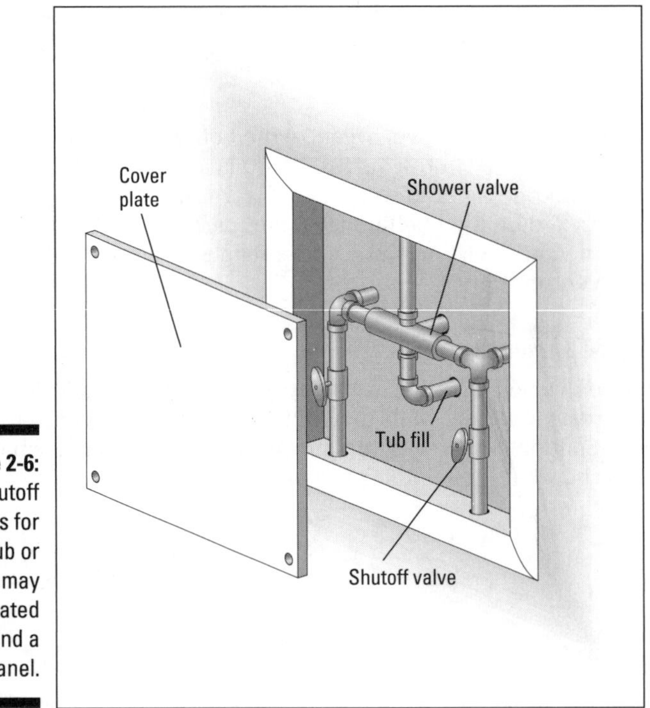

Figure 2-6: Shutoff valves for the tub or shower may be located behind a panel.

If you either can't find the panel we describe here or you find the panel but then don't see any valves behind it, you haven't reached the end of your rope just yet. Many newer bathtubs and showers have shutoff valves installed behind the shower handle. Single-handle shower valves (in which you control the hot and cold water with one handle) have a cover plate sealing off a hole large enough that you can reach the shutoff valves from the bathroom side of the wall. If you remove the handle and cover, you may find shutoff valves (like the ones shown in Figure 2-7) installed on the hot and cold lines just before they enter the valve body. If you have a double-handled shower valve, chances are you don't have built-in shutoff valves in this type fixture.

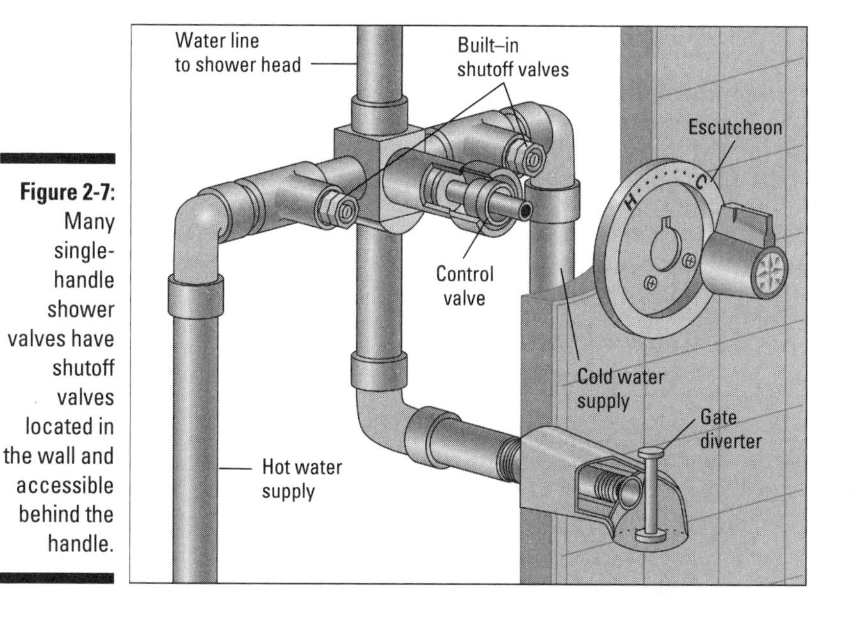

Figure 2-7: Many single-handle shower valves have shutoff valves located in the wall and accessible behind the handle.

In the kitchen

The kitchen is another big place for water usage in any home. Depending on the appliances you have installed in your kitchen, you may have water running to one or two sinks, a dishwasher, and, probably, your freezer's icemaker. The following sections give you all the help you need to find the shutoff valves for these popular kitchen features so that you can get on with the task of repairing them.

The sink

The shutoff valves to the kitchen sink are located inside the base cabinet directly underneath the sink. The supply lines may either come out of the wall or emerge from the floor.

If the supply lines come out of the wall, you'll find a shutoff valve on the hot line and the cold line. If the lines come out of the floor and you can't find shutoff valves on these lines, you may find a pair of valves in the basement, under a sink.

The dishwasher

If your dishwasher is next to the sink or in the same section of base cabinets as the sink, the shutoff valve is probably under the sink. Often, when a dishwasher is located near or next to the kitchen sink, the pipe leading to the dishwasher is connected to the pipe that supplies hot water to the sink. In this case, two shutoff valves control the hot water flow: One valve goes straight up to the hot water connection to the faucet; the other valve goes off to the side of the dishwasher.

If the dishwasher isn't located next to the sink, the water supply comes out of the wall or the floor. The shutoff valve is located under the dishwasher. You can get to this area by removing the kick plate cover of the dishwasher, which should pull off easily. If you have trouble removing the kick plate cover, try lifting it slightly and then pulling it out. (See your owner's manual for specific instructions on how to remove this panel.) You should find the shutoff valve under the dishwasher, usually in the front, close to the water supply line running into the unit, as shown in Figure 2-8.

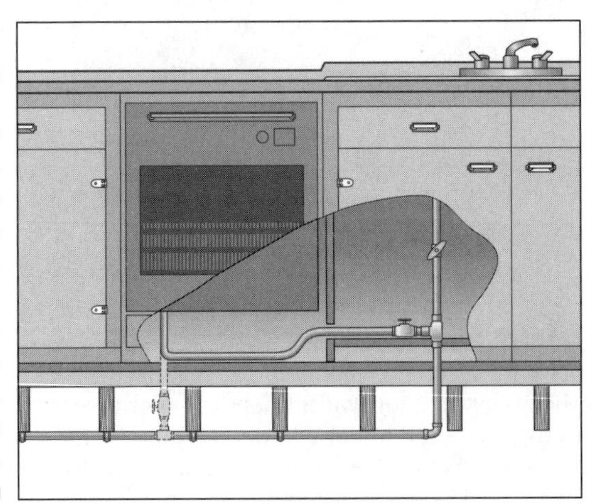

Figure 2-8: The shutoff valve for your dishwasher may be located underneath the appliance, as shown here.

The icemaker

Your freezer's icemaker has its own water supply, which runs through a ¼-inch plastic or copper tube. In most cases, the tube leads to a larger pipe, where it is connected to a shutoff valve. Follow the tubing to the larger pipe and look for the valve.

If you don't see the valve, your icemaker may work a little differently. The water supply still runs through a ¼-inch tube. But that tube instead may lead to a larger pipe that has a clamp-on device, called a *saddle.* The saddle has a shutoff valve on it, which controls the supply of water.

In the laundry room

The laundry room is another major place in your house where you use water. Your washing machine gets its water supply from something called a *hose bib,* which is hooked up inside a laundry box. This box contains the water supply and the drain pipe for the washer, and is usually located in the wall just behind the washer. Instead of shutoff valves, you use the handles on the hot and cold hose bibs to turn the water on and off.

In a house without a laundry box, the hot and cold water lines are usually part of a laundry tub, and the faucets are used to connect the water supply pipes to the washing machine.

In the great outdoors

If your house has outdoor faucets, called a *bib faucet* or *sillcock,* you can find the shutoff valve, called a *bib valve,* in your basement or crawl space (located in the pipe leading to the faucet). Close the bib valve, and then open the bib faucet to allow the water between the bib valve and the bib faucet to run out.

Letting the water run out of the pipes leading to your outdoor faucets prevents your pipes from freezing in low temperatures.

Some bib faucets are controlled by a valve that is designed to make draining the water out of the exterior portion of the pipe easier. This valve has a small, curled screw — called a waste screw — on its side (see Figure 2-9). When you remove the screw, air can enter the pipe and expedite the drainage. Turn the valve off, open the outside bib faucet, and then unscrew the cap on the side of the stop and waste valve. Doing this will assure that all of the water drains out of the pipe. Then just replace the cap.

Many houses have *frost-proof bib faucets,* for which the actual shutoff valves are located well inside the house to prevent the faucets from freezing in the winter months. (See Figure 2-9.) If you have frost-proof bib faucets (also called a *freezeless sillcock*), you don't need to shut off the water at the valves inside the basement or crawlspace during the winter. Instead, these faucets have a thick pipe leading into the house. If you need to turn off the water supply for repair purposes, look for the shutoff valve at the inside end of the pipe, where it connects to the water supply pipe.

Never leave a hose attached to the frost-proof bib during freezing weather. Water won't be allowed to drain out, and can cause serious damage to frost-proof bibs.

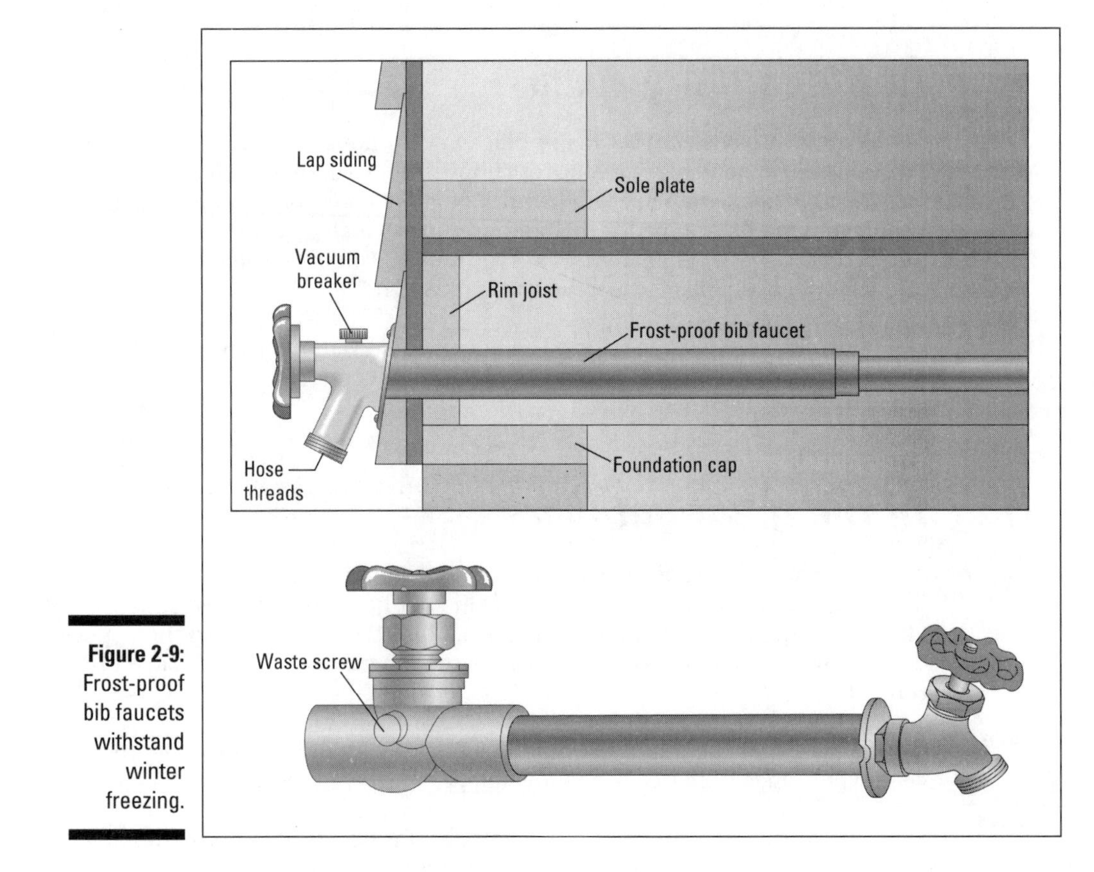

Figure 2-9: Frost-proof bib faucets withstand winter freezing.

Chapter 3

Materials, Supplies, and Tools

In This Chapter

- Getting a primer on replacement parts
- Choosing the right valves and other supplies
- Figuring out what kind of pipes and fittings you need
- Getting your hands on all of the right plumbing tools

A hardware store clerk can always tell when a customer has a plumbing project. The obvious telltale sign is the distraught look of someone carrying a rusted pipe or odd-shaped piece of chrome. But it's the dazed and glazed look or mindless stare that tells it all — the homeowner is clueless and fearful knowing what to buy to solve his or her plumbing problem.

You can avoid being that hopeless soul by taking a look at the materials, supplies, and tolls that we've gathered in this chapter. With a basic understanding of what this stuff is and what it does, you can walk down the aisles with an air of confidence. You still may be clueless, but you'll be armed with the vocabulary — and that's half the battle.

The larger the plumbing department, the greater the room for confusion. If you're lucky, you can find a hardware store, home center, or plumbing supply house that has plenty of in-store displays to demonstrate how plumbing parts and materials go together. If you're *really* lucky, you'll land in a store where the folks behind the counter or in the aisle have some real-life plumbing expertise.

Finding Plumbing Replacement Parts

Many times, your visit to the plumbing department is to find small replacement parts, rather than large, new products. Take a look-see at these most popular replacement parts for your plumbing appliances and fixtures so that, when the need arises, you know just what to ask for. This stuff may seem ho-hum, but knowing what you're looking for can save you countless return trips to the store.

- **Washers:** A variety of washers are used in plumbing applications and generally help to seal a temporary joint. Probably the most common washer is used inside the *female end* (the end that screws onto the faucet) of a garden hose. Washers are also used inside a faucet to seal the joint between the smooth *valve seat* that's inside the faucet and the valve stem, to prevent the flow of water. When you open a faucet, the washer is raised above the valve seat, which allows the flow of water.
- **Aerator:** An *aerator* is a small system of screens and a baffle that is screwed into the end of many kitchen and bathroom faucets. Aerators introduce air into the flow of water and make it appear foamy. Aerators can become clogged with debris, so clean them regularly. To remove, simply unscrew the unit, clean, and reinstall.
- **Valve seat:** A valve seat is present in the inside bottom of most faucets. These small (about the size of a dime), donut-shaped, brass seals are threaded on the bottom and are screwed into place with a special *seat removing tool* that's not unlike a large Allen wrench — see Figure 3-1. Because they are made from brass, which is very soft, valve seats deteriorate in time, causing the faucet to drip.
- **Valve stem:** A valve stem is located under the handle of both hot and cold water faucets. On the bottom of every valve stem is a washer that wears out over time. If your faucet is leaking, the first thing to replace is this washer. You can also find an o-ring or stem packing (which looks like string wrapped around the stem) that's designed to prevent water from leaking around the base of the handle (see Figure 3-1). Either the o-ring or stem packing should be replaced if water leaks from the base of the faucet stem.
- **Washerless faucet parts:** Washerless faucets don't have valve seats, washers, or valve stems. Instead they have a ball mechanism — the most common type being a kitchen faucet with a single lever handle. If this type of faucet starts to drip, the internal o-rings and/or ball mechanism need to be replaced. Replacement kits are widely available in home improvement and plumbing supply shops. Some washerless faucets have a cartridge-type mechanism that you can replace. Flip to Chapter 15 for more details on faucets.
- **Toilet float:** A toilet float, shown in Figure 3-2, is a mechanism inside the toilet tank that turns on the water when you flush and turns off the water after the tank has refilled. Some toilets have a ball float — about the size of a softball, on the end of a metal rod; other designs use a can-shaped cylinder that moves up and down according to the water level in the tank. An overflow tube inside the tank prevents the tank from overfilling.
- **Toilet flapper:** A toilet flapper is located in the bottom of the toilet tank. When you push the flush lever down, a small chain lifts the flapper up and allows the water in the tank to flow and flush the toilet.

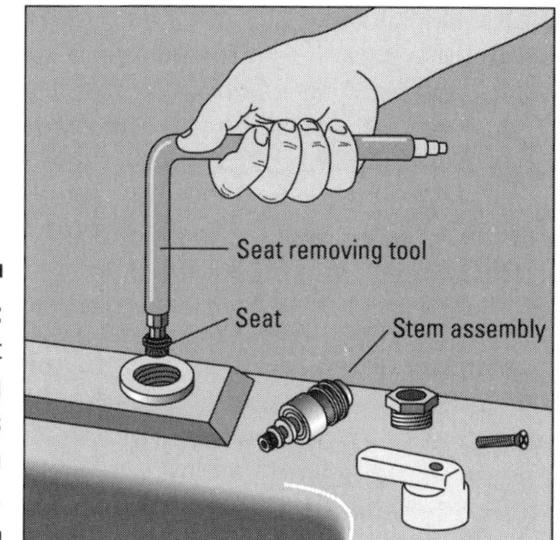

Figure 3-1: Faucet showing valve seats and stem assembly.

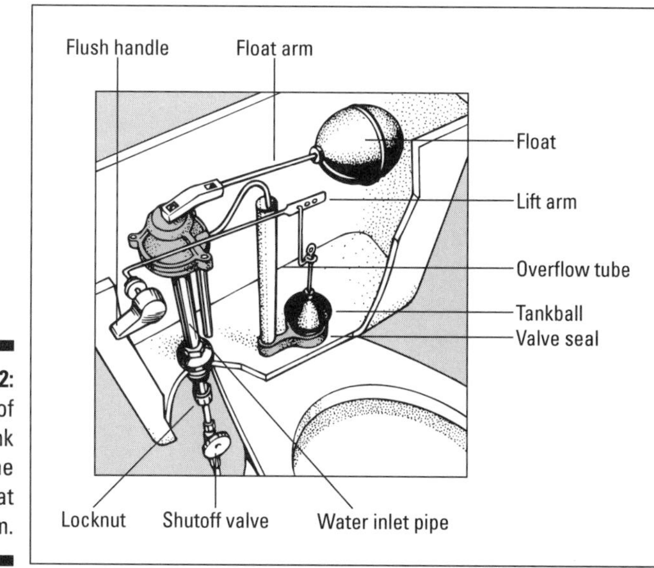

Figure 3-2: Interior of toilet tank showing the toilet float mechanism.

Understanding Common Plumbing Supplies

Just like any good cook has a kitchen full of supplies to create new and exciting entrees, a do-it-yourself plumber needs to develop a stash of stuff to work with pipes and fixtures. Here's a rundown of the basic materials to have on hand.

- **Plumber's putty:** This material looks like modeling clay and is designed to stay soft and semi-flexible for years. Use it for making a seal between plumbing fixtures and in areas where there is no water pressure. When you install a new faucet, for example, you apply plumber's putty under the faucet where the faucet meets the top of the sink — and helps to prevent water from seeping under the faucet and into the sink cabinet below. Other uses for plumber's putty include sealing drains in sinks, bathtubs, and shower stalls; sealing frames around sinks; and sealing other fixtures and bowls.
- **Pipe joint compound:** Pipe joint compound is sometimes referred to as *pipe dope* and is used to seal the joint between a threaded fitting and steel pipe. Always use pipe joint compound when working with gas lines; many people also use it when working with galvanized steel water lines. Pipe joint compound is typically painted onto the threads cut into the end of the pipe that's called the *male end,* as shown in Figure 3-3.

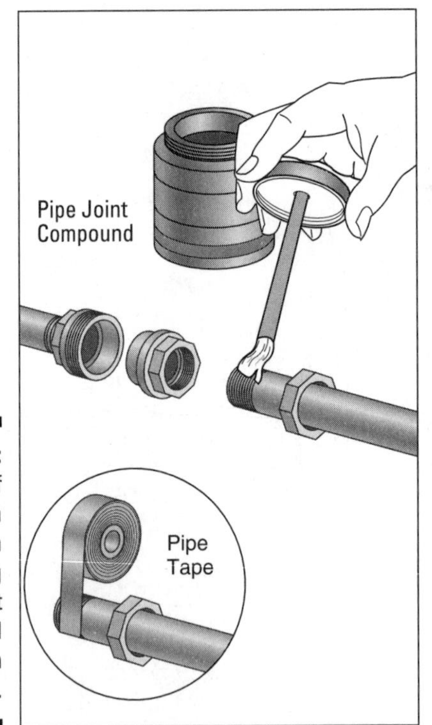

Figure 3-3: A section of pipe with a union showing various joint compound and Teflon tape.

- **Teflon tape:** This tape is a substitute for pipe joint compound and is used to seal the threads on steel pipe — see Figure 3-3. Use white Teflon tape for sealing pipe threads; yellow Teflon tape, which is much thicker, for sealing pipe threads when working with gas lines.

- **CPVC cleaner:** CPVC cleaner is used for cleaning CPVC pipe prior to joining it together with adhesive (see the "Buying Drainage Pipes and Fittings" section, later in this chapter, to find out more about CPVC pipes). This purple-colored liquid not only cleans PVC pipe; it also slightly softens the pipe and makes the adhesive work better.

  Be careful using PVC cleaner — it stains wherever it lands.

- **CPVC cement:** CPVC cement is used to fuse joints in CPVC pipe and fittings.

- **ABS cement:** ABS cement is used to fuse joints on black ABS pipe and fittings (see the "Buying Drainage Pipes and Fittings" section, later in this chapter, to find out more about ABS pipes and fittings). A cleaner is not required prior to using ABS cement, but the pipe and fittings should be wiped with a clean cloth.

  Never use PVC cement on ABS pipe — it's incompatible.

- **Solder:** Solder (pronounced SOD-der), is used for sealing joints when working with copper pipe. The pipe and fitting are heated with a propane torch until they're hot enough to melt the solder in a process called *soldering*. Solder is sold in rolls — most commonly 1 pound — and is made from various combinations of pure lead and tin. 50/50 and 60/40 are good types of solder to use for plumbing joints. The first number is lead and the second is the amount of tin in this particular solder. The addition of tin helps the solder stick to the copper.

- **Leadless solder:** Leadless solder is used to join water supply pipes and is required by many building codes because standard lead-based solder can leach lead into the water standing in the pipe. Your best bet is to use leadless solder for all plumbing projects, not just for *potable* (drinking) water supply.

- **Flux:** Flux is used to clean the copper pipe and helps the solder adhere better, and is available in paste or liquid form, but most plumbers prefer the paste type. Flux must be used when soldering copper pipe; it is applied with as small brush to both the end of the pipe end inside the fitting — see Figure 3-4.

  TIP: Always wear inexpensive jersey gloves when working with solder and flux. The gloves eventually get eaten up (so use a cheap pair), but they'll protect your hands.

- **Steel wool:** Use steel wool to clean the ends of copper pipe and the inside of copper fittings, prior to applying flux and soldering.

- **Emery cloth:** Emery cloth is a cloth-backed sandpaper that's used to clean the ends of copper pipe and the inside of copper fittings prior to applying flux and soldering.

- **Copper cleaning brush:** This is a small wire brush used to clean the inside of copper fittings — it cleans a fitting faster than an emery cloth does. Copper cleaning brushes are sold in sizes to fit inside ½-inch and ¾-inch fittings.

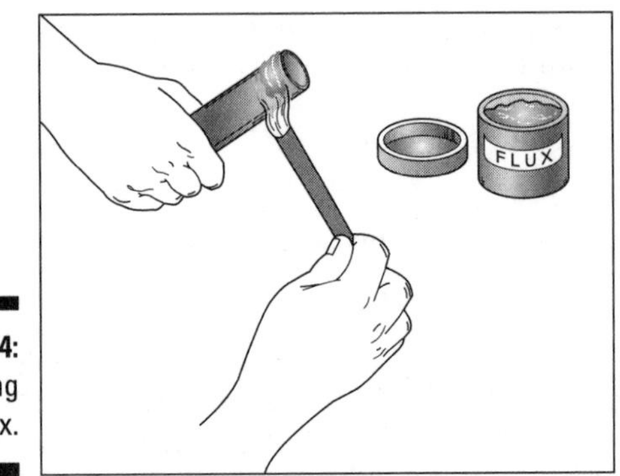

Figure 3-4: Applying flux.

- **Propane:** Propane is a liquid petroleum product that is used in a hand-held torch, which is used to heat copper pipe and fittings, and to melt solder to seal the joint.
- **Wax bowl ring:** These large, donut-shaped rings are inexpensive and made from a special wax compound that seals out gasses and water when installing a toilet.

  TIP

  Always install a new wax bowl ring whenever you're working with a toilet and the original seal has been disturbed.
- **Pipe thread cutting oil:** This light oil is used when threading the ends of steel pipe. Pipe thread cutting oil helps to lubricate the thread cutter and to wash away the steel particles that are created when the pipe threads are cut.
- **Old rags:** Plumbing is a messy job, so keep plenty of old rags on hand. These come in handy for general clean up and for wiping the excess solder from a soldered copper joint.

Finding the Right Water Supply Pipe

Water supply pipes are an important part of plumbing projects, and they're available in a variety of materials, including copper, iron, and plastic. The wide acceptance of plastic pipe by most building departments has been a big boom to the do-it-yourself plumbing industry, because plastic pipe is easy to cut and glue together. Most homes older than 20 years most likely have pipes made of iron or copper.

Copper pipes

Copper pipe, used for water supply lines throughout most homes, is probably the most common type of pipe in use today. It is widely available in two basic types: rigid and flexible. Both types come in several wall thickness. This section gives you a basic rundown of copper pipe and its uses.

Rigid copper pipes

Rigid copper pipe is used extensively throughout modern homes for both hot and cold water supply lines. This pipe is widely sold in both ½- and ¾-inch diameters and in lengths of 8 to 20 feet.

Most suppliers carry several grades of rigid copper pipe, each having a different wall thickness — the most readily available are types L and M. Type K is a heavy-duty grade and is usually sold only in plumbing supply outlets.

Rigid copper pipes obviously don't bend, so a variety of fittings, shown in Figure 3-5, are available to help you make the pipe go where you want it to go:

- **Elbow:** Use an elbow to make copper pipe turn at an angle. Elbows (commonly called *els* in the trade) come in 90-degree, 45-degree, and 22½-degree angles.
- **Tee:** A tee allows you to run another copper line off an existing line.
- **Straight connector or coupling:** Use these when you want to continue a straight run of consecutive copper pipes.
- **Cap:** Use copper caps to end a line of copper pipe.

All fittings for rigid copper pipe are soldered in place. (See the "Understanding Common Plumbing Supplies" section, earlier in this chapter, for an explanation of soldering.) This involves the use of solder, flux (so that the solder will stick), and heat (usually from propane torch).

Flexible copper pipes

Flexible copper pipe is sold in coils of various lengths, and is also available in two grades, K and L. Most stores stock the L grade.

Common sizes of flexible copper piping include ⅜-inch, ½-inch, and ¾-inch diameters. The ⅜-inch size is commonly used for hooking up a water supply line to dishwashers and some freezer icemakers. Flexible copper pipe and tubing are commonly joined with *compression fittings* rather than soldering.

When you buy compression fittings for flexible copper pipes and tubing, the fitting comes with a single *compression ring.* When you tighten a compression fitting with a wrench, the compression ring seals the joint — but not always. If the joint leaks even after you tighten it, the compression ring may have been damaged when you tightened it. Purchase a few extra compression rings, so that you can remove the damaged one and try again.

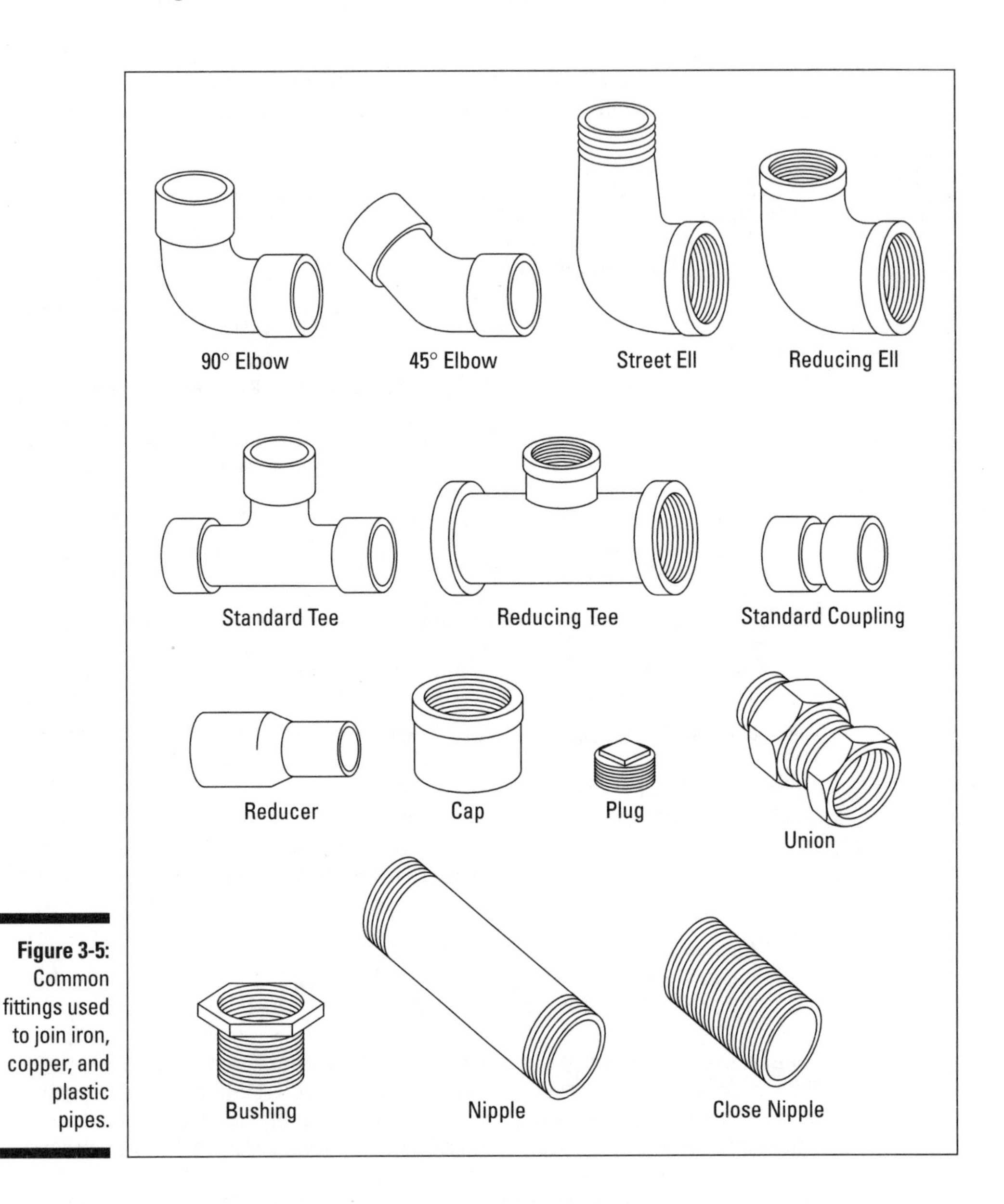

Figure 3-5: Common fittings used to join iron, copper, and plastic pipes.

Galvanized iron pipes

Years ago, galvanized iron pipe was the only type of pipe that was widely available for water supply lines. Today, other types of pipe, such as copper and plastic, are beginning to replace galvanized iron, because they require less labor to install. Depending on the age of your house and the part of the country that you live in, you may or may not have galvanized iron pipes. If you have galvanized iron water pipes, you'll find two basic pipe sizes in your walls: ½-inch diameter and ¾-inch diameter.

If you aren't sure whether your pipes are copper or iron, try to stick a magnet to the pipe. If the magnet sticks, the pipe is iron; if it falls, it's copper.

Because the ends of each pipe are threaded, you need a special thread cutting tool. You can rent this tool, along with a pipe vise to hold the pipe while you cut the threads.

Black iron pipe is used only for gas piping. Water heaters, furnaces, stoves, and other appliances that use gas are plumbed with black iron pipe. This pipe is available in the same sizes and has the same fitting configurations as galvanized iron pipe.

When you need to join galvanized iron pipes together, you use the same kinds of fittings you use for copper pipes — elbows, tees, straight connectors, and caps (see "Rigid copper pipes" earlier in this chapter for more information on these specific fittings). In addition to these fittings, you also use a fitting called a *union* to join iron pipes. Unions allow two sections of iron pipe to be joined or taken apart. Because iron pipe is threaded at both ends, you can only thread one end of the pipe into a fitting at a time. Pipes thread into fittings by turning them in a clockwise direction. As you thread the pipe into a fitting, the end that's going into the fitting is turning in a clockwise direction, but the other end of the pipe is turning counterclockwise. So you must thread pipe into fittings and fittings onto pipes in a sequential manner, which means that you have to plan ahead. (See Chapter 9 for tips on repairing iron pipes. That chapter also shows you how to convert from iron pipe to easily worked plastic pipe.)

Plastic pipes

Depending on your local building codes, you may or may not be able to use plastic for water supply pipes in your home. Plastic pipe is used in some parts of the country for water supply lines, but many local building codes prevent its use. The most common use for plastic pipe is in waste, vent, and drain lines (see Chapter 2). The second most popular use of plastic pipe is for underground sprinkling systems for lawn irrigation. Check with your local building department about their specific requirements.

Two types of plastic water supply pipe are commonly used today: Chlorinated polyvinyl chloride (CPVC) and polybutylene (PV) flexible tubing. Both kinds of pipe are available in ¼-inch, ½-inch, ¾-inch, and 1-inch diameters.

Plastic pipe is easy to work with, because it can be cut to length with a fine-toothed saw. Joints for plastic pipes include elbows, tees, and caps (refer to Figure 3-5). Seal all of the joints in plastic pipes with a liquid adhesive. Be sure to clean the joints for CPVC pipe first with a special cleaner before applying a coat of adhesive. (You can find more information on these cleaners and adhesives in "Understanding Common Plumbing Supplies" section, earlier in this chapter.)

PV tubing is joined with compression fittings. This type of tubing is joined with compression-type fittings or standard flared-type fittings that are used to join flexible copper pipe.

When you purchase any type of supply tube, always buy a longer tube than you need. You can cut copper and plastic tubes to the length you need.

Buying Drainage Pipes and Fittings

Pipes that carries water out of the drainage system of your house look different from the water supply lines that bring it into the house. Fittings designed for drainage have to make gentle turns. Standard elbows used to join pressure pipes turn abruptly at 90 degrees, but a drainage elbow has a gradual bend. Drainage tee- and wye-fittings have a 45-degree angle.

Cast iron

Cast iron pipe, commonly used for drain lines in older homes, has now largely been replaced by plastic drain lines. Joining cast iron drain lines involves molten lead and plenty of messiness. If you have to mess with cast iron drainage pipes, we strongly recommend you call in the pros. This pipe is heavy, difficult to work with, and requires very specialized tools.

Do it once — do it right

If you're planning to replace some old galvanized iron pipe with plastic, find out if plastic is allowed. You may think that it won't make any difference because the inspector will never know. He won't until you (or the person who inherits your house) sells the property. When the buyer has the house inspected and materials are found that aren't allowed, they'll have to be replaced.

An old-time remodeler gave us some great advice the first time we added a new bathroom. He told us to get a building permit and then have the plumbing plan approved by the city inspector. The inspector changed a couple of things, but when he inspected the plumbing it passed with no sweat. It's the best way to avoid problems.

WARNING!

Supply tubes

Supply tubes connect a faucet or toilet to a shut-off valve and can be used for hot or cold water. Supply tubes are most commonly ⅜ inch in diameter but are also available in ½-inch diameter. Supply tubes come in three basic forms: rigid copper (most commonly chrome), CPVC plastic, and flexible copper. Flexible copper is easy to work with and the most dependable of the three materials. But the CPVC plastic supply tubes are by far the easiest to install.

All supply tubes are not created equal. The shape of the end of the tubing used to supply water to a toilet is slightly different than that of the tubing that supplies water to a sink. Read the description on the package of tubing — it should clearly state whether the tube should be used for a toilet or sink, or it may be a universal type suitable for both.

Plastic pipes

Plastic drainage pipes and fittings have revolutionized the plumbing industry. They are much lighter than cast iron and a lot easier to work with. Most building codes will allow plastic drainage systems. But before you begin any plumbing project, check your local building codes to determine exactly which types of pipes are allowed in your area. (See Chapter 1 for more information on building codes and getting a building permit.)

Two basic kinds of drainage pipe are commonly used today: polyvinyl chloride (PVC) and acryaonitrile-butadiene-styrene (ABS). Both type pipes are suitable for drainage. This type of pipe is available in several wall thicknesses; schedule 40 is the most widely available. Common diameters for these pipes include 1½-inch, 2-inch, 3-inch, and 4-inch diameters. You can also find corresponding plastic fittings to match the pipes produced in cast iron, which are helpful if you want to replace iron with plastic in your house. (See Figure 3-6.)

When you're choosing fittings for pipes under 2 inches in diameter, make sure you choose a drainage fitting and not one designed for supply piping.

When you're buying adhesive for ABS or PVC plastic pipe, always purchase the type of adhesive recommended by the pipe manufacturer. Although so-called "universal" adhesives are available, most local building codes require ABS adhesive for ABS pipe and PVC adhesive for joining PVC pipe.

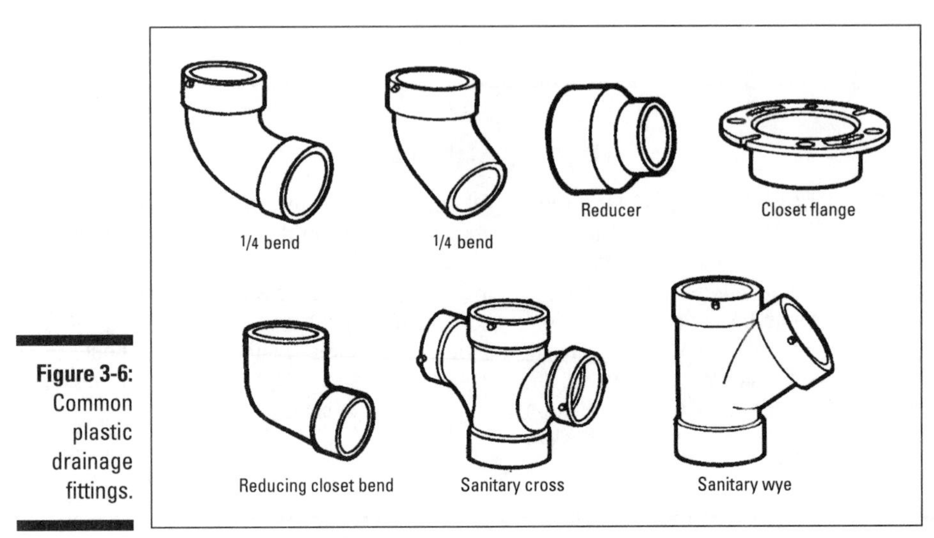

Figure 3-6: Common plastic drainage fittings.

Illustration provided by Genova Products, Inc.

Stocking up on Valves

Valves control the flow of water in pipes. Your home has a valve on the incoming water line so you can stop the flow of water into your home (see Chapter 2).

Some valves have a round knob on top — which must be turned to open or close the valve — and other valves have a single lever that is pushed or pulled to open or close the valve. Valves are used for all types of plumbing pipe and are available with threaded ends (for used with galvanized steel pipe) or smooth ends (for use with soldered copper joints). You can also find valves with plastic bodies, for use with adhesive in a plastic pipe application.

When you're buying valves, be sure you buy the proper kind. Some valves have threaded female ends that accept the threaded end of a pipe. Valves are also available to attach to copper and plastic piping — these have unthreaded female ends designed to be soldered or solvent-welded to the pipe. Still others have a compression-type female ends that clamp down on the pipe when the compression nut is tightened.

Many different varieties of valves exist, as evidenced by the following list, in which we introduce the kinds of valves that you're most likely to encounter. See Figure 3-7 for examples of these valves.

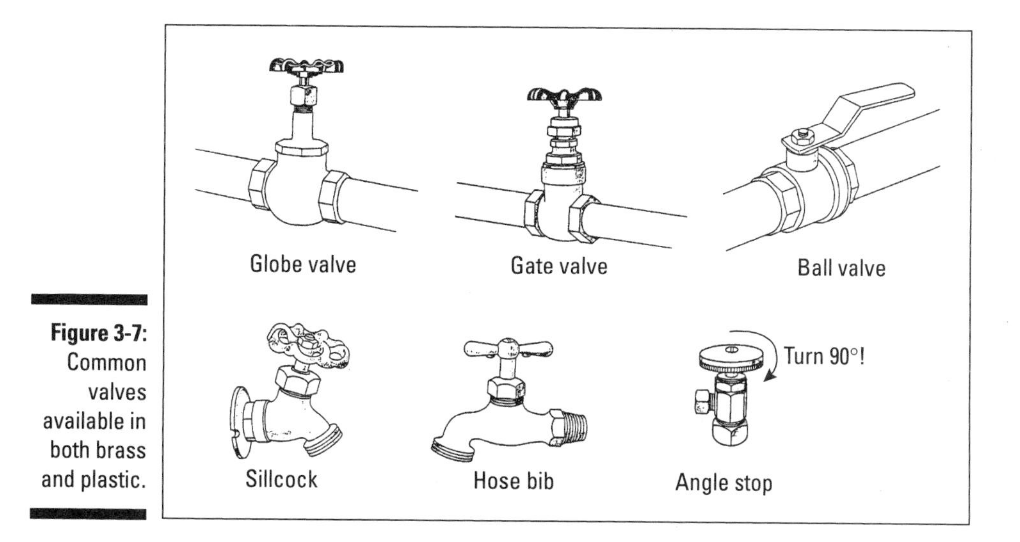

Figure 3-7: Common valves available in both brass and plastic.

- **Gate valve:** Gate valves allow the full flow of water through the valve. This valve is the best kind to use in the main shutoff valve and in pipes supplying cold water to the water heater and other devices where water flow is important. Gate valves don't have rubber compression gaskets; instead, they rely on the close fit of the parts to stop the water. Turning the handle counterclockwise raises the wedge-shaped gate, allowing the water to flow.
- **Ball valve:** Ball valves can provide full flow of water. They are opened and closed by a lever. Inside the valve is a ball with a hole through it. When the hole in the ball is aligned parallel with the pipe, water flows through the valve. When the ball is turned with the hole across the pipe, the water is blocked. The lever is aligned with the hole in the ball. To close a ball valve, push the lever perpendicular to the pipe; to open it, push the lever parallel with the pipe.
- **Globe valve:** Globe valves, which are less expensive than gate valves, are a good choice for general water control where full flow is not necessary. Globe valves have a rubber compression washer that is pushed against a seat to squeeze the water flow.
- **Stop valve:** Stop valves are a variation of globe valves. They are designed to control water running to sinks and toilets. Stop valves come in two main varieties: angle and straight stops. *Angle stops* are designed to control the water flow and change the direction of the flow 90 degrees. Angle stops are mounted on the end of pipes under sinks and toilets when the supply pipes come out of the wall. *Straight stops* are designed to be mounted on pipes that come out of the floor.

Be sure you choose a valve with the right kind of outlet. The valve is designed to attach to the pipe coming out of the wall or floor and to the riser tube going to the fixture. If, for example, you want a stop valve to attach to a ½-inch galvanized pipe and to a ⅜-inch riser tube, purchase a valve with a ½-inch female threaded end and a ⅜-inch outlet to accept the tubing.

- **Hose bib:** A hose bib looks a lot like a miniature faucet. It has a male-threaded end that is screwed into a coupling threaded to the end of a pipe. Hose bibs are commonly used to control water running to washing machines.
- **Sillcock valves:** Sillcock valves are used outside the house, on the faucets that you most likely use with a garden hose (called a *bib faucet*). For houses located in cold winter climates, a freeze-proof version is available (see Chapter 2).

Using Plumbing Tools

When you're embarking on a new plumbing project, you have the upper hand if that hand is holding a tool designed for the job. Sure, a plain old screwdriver can tackle many of the jobs you need to do. We're not talking about pouring hundreds of dollars into outfitting your toolbox with tools. What we do suggest is that you begin to acquire tools as you need them, based on the suggestions we offer in this section. That way, you'll be sure to slowly acquire the tools you need, and you'll have them for a lifetime.

Basic woodworking tools

As with any repair work you do around the house, you need a set of basic tools. Woodworking tools are a good place to start, because if you decide to replace a kitchen sink, for example (see Chapter 13), you'll be cutting and drilling holes in wood, driving screws, tightening bolts, and doing all kinds of jobs that at first glance may seem unrelated to plumbing.

These are the common woodworking tools you should have in your toolbox:

- ⅜-inch variable-speed, reversible drill with a variety of drill bits
- Utility knife, claw hammer, and assorted screwdrivers
- Slip-joint pliers: A tool with curved-toothed jaws and a hinge that adjusts or "slips," making the jaw opening wide or narrow.
- Locking pliers (also known as Vise-Grips): Pliers with short jaws that may be opened and set to a specific size by turning a jaw-adjustment screw in the back of the handle. When the handles are squeezed the jaws are clamped together.

- Adjustable (Crescent) wrench: This wrench has a long, steel handle and parallel jaws that open and close by adjusting a screw gear.
- Staple gun and hacksaw
- Carpenter's level: A straight-edge tool that has a series of glass tubes, each of which contain liquid and a bubble of air. It shows that a surface is *level* (horizontal) or *plumb* (vertical) when the bubble in a single tube is framed between marks on the glass.

- Torpedo level: This tool is a shorter 8- to 10-inch version of a carpenter's level. It's useful for plumbing applications, because it fits into restricted areas.
- Metal files: These are used to shape and sharpen wood and metal surfaces. You want to have a flat and half-round type with shallow grooves that form teeth.
- Set of Allen wrenches: Short, L-shaped metal bars that are designed to turn screws or bolts which have hexagonal sockets in their heads.
- Retractable tape measure

Tools for measuring

Accurate measuring is essential when installing new water or drain lines. As we point out in the "Finding the Right Water Supply Pipe" section, earlier in this chapter, you need to know the diameter and length of the pipe you need. A retractable tape measure can do most of the measuring you'll need to do — the hook on the end of the retractable tape measure catches on the end of the pipe, making it easy for you to mark a cutting point.

For more precise measurements, use a measuring device called a *steel rule,* which has both English and metric graduations. Steel rules come in lengths of 12 to 48 inches. (The shorter length is easier to use for measuring plumbing-related pieces.) Place one end of the rule against an inside edge of the pipe, and read the size at the opposite edge. This measurement may have little relationship to the pipe's *nominal size,* or outside diameter.

Wrenches and pliers

A number of tools have been designed to help you work efficiently with different kinds of pipe. Other special tools make it easier to remove and install plumbing fixtures and associated components, such as water supply lines and drain lines.

Some of these tools, like a pipe wrench, monkey wrench, basin wrench, and groove-joint pliers, are the minimum you need for basic plumbing repairs. If you have these tools on hand, you'll be ready if you ever have an unexpected plumbing failure. (We don't know of anyone who has planned failures.) So you should have them on hand even if you don't plan any big plumbing projects in the near future. Besides, you'll be able to use many of these tools for other jobs around the house.

Other specialty plumbing tools are available (see the following bulleted list), and you may need them for one-time use. So don't go out and buy them unless the need arises. Better yet, figure out if you may be able to rent them.

- **Pipe wrench:** This wrench, shown in Figure 3-8, has serrated teeth on the jaws, which are designed to grip and turn metal pipe. One jaw is adjustable, with a knurled knob. This jaw is spring-loaded and angled slightly, which allows you to release the grip and reposition the wrench without changing the adjustment. Pulling the handle against the open side of the jaws causes them to tighten against the pipe. Pipe wrenches come in different sizes, with 8-inch, 10-inch, and 14-inch being the most useful.

  Don't use pipe wrenches on copper or plastic pipe. You don't have to turn a copper or plastic pipe since it is soldered or solvent-welded (respectively) to the fitting. A pipe wrench can also crush the pipe if a lot of pressure is applied to the pipe with the wrench.

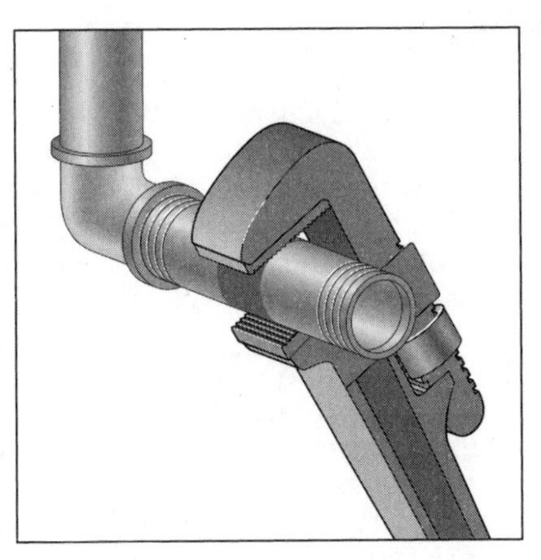

Figure 3-8: After the teeth on the jaws of a pipe wrench grip a pipe, they won't let go until you release the handle.

- **Monkey wrench:** Similar to an adjustable (Crescent) wrench in purpose, the jaws of a monkey wrench are parallel to each other and set 90 degrees to the handle.
- **Spud or trap wrench:** This wrench is an adjustable one that's designed for handling drain-trap and sink-strainer fittings (see Chapter 13). The jaws are parallel and have a wide adjustment so they can grip the large diameter lock ring that holds a sink strainer assembly in place. You set the jaws to size, and then lock them in place by tightening a wing nut on the body of the wrench.

- **Strap wrench:** This tool has a canvas webbing that wraps around pipes or plumbing fittings and allows you to tighten or loosen them without marring the finish.
- **Chain wrench:** This wrench is designed to handle a wide range of pipe sizes. The chain operates in the same fashion as a strap wrench. The chain wraps around the pipe and is hooked onto the handle. Unless you plan to work on large-diameter galvanized pipes (and we don't recommend this), you can certainly get by without this tool.

- **Basin wrench:** One of the hardest nuts to reach is the one that holds a sink faucet in place. The basin wrench, shown in Figure 3-9, fills the need. It has a long handle that reaches the nut. The jaw is hinged and repositions itself after each turn. Spend a little extra money and buy one that has a reversible jaw. You'll still have to crawl in the cabinet under the sink to see the nut you're after, but at least you won't need a 4-foot arm with six wrists to get the job done.
- **Plastic nut basin wrench:** Many faucets have easy-to-tighten plastic mounting nuts, but they're still hard to reach. So the plastic nut basin wrench is about 11 inches long and is designed to reach and tighten these nuts. This metal tool has notched ends, which self-center on 2-, 3-, 4-, and 6-tab nuts, and they also fit metal hex nuts.
- **Groove-joint pliers:** In larger sizes, these pliers are used for holding pipes. With long-reaching parallel jaws, you can also tighten drains and use them for a multitude of other jobs.
- **Faucet spanner:** The faucet spanner is a flat wrench, usually stamped from heavy metal, used for installing faucets. You may get one when you buy a new faucet. These wrenches have a variety of hex or square holes punched in the wrench, sized specifically to fit packing nuts and other fittings on faucets.

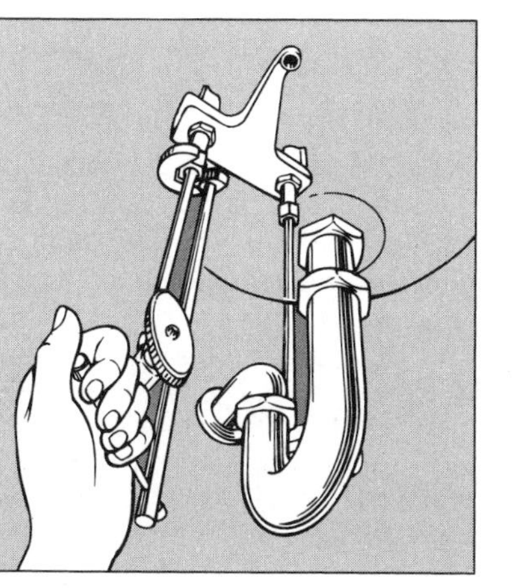

Figure 3-9: A basin wrench makes a tough job a little easier after you get the hang of using it.

Pipe clamps

Pipe clamps come in a variety of shapes and sizes. You use them to make temporary emergency repairs on a pipe that may have frozen and burst or sprung a pinhole leak due to corrosion. Place a rubber or plastic pad over the leak, and then install the clamp over the pad and tighten — see Figure 3-10. The clamp should extend at least one inch beyond the leak.

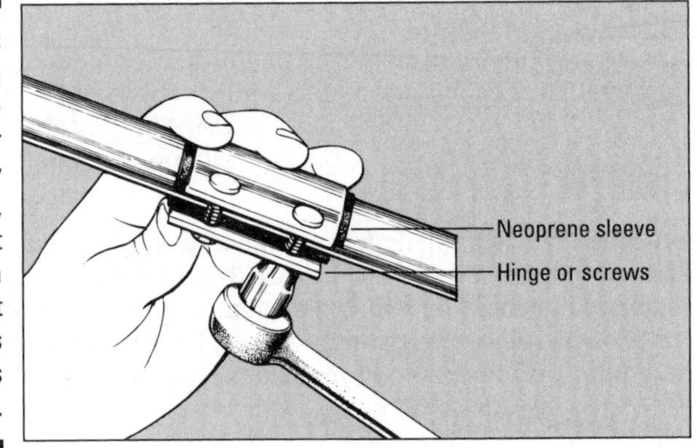

Figure 3-10: After using a pipe clamp for emergency repairs, don't forget to make a permanent repair as soon as you can.

Pipe cutting and bending tools

Some tools are made specifically to bend pipes, cut pipes, and dress the cut end (see Chapter 4). When you need one of these tools, no substitution will do. Consider renting these tools if your pipe-cutting project is a small one. But if you'll be involved in more than one pipe-cutting job, consider investing in the tools. You'll enjoy a smug satisfaction when you can talk authoritatively to friends and coworkers about your pipe reamer, knowing they don't have a clue what you're saying.

- **Tubing cutter:** Using a round cutting wheel, this tool makes neat, straight cuts on copper tubing and plastic pipe or tubing. Tube cutters come in sizes to fit tubing of different diameters. The most convenient size for homeowners is a model that will cut tubing from 3⁄16-inch to 1⅛-inch in diameter. Some are fitted with a built-in *reamer* to remove burrs inside the pipe after cutting. A *midget tube cutter* is designed for use in tight, tiny spaces. It doesn't have a handle, but large, knurled, feed-screw knobs make it fairly easy to use.

GREAT GADGETS

Tools to rent

The following tools are available from most tool rental shops. Rent them as needed — you won't use them enough to make them worth purchasing.

- **Tools for drilling large diameter holes:** When you install new plumbing, you'll probably need to bore some holes in studs or floor joists. For water supply pipes, a 1 ¼-inch spade bit works just fine. This is an inexpensive drill bit made from a flat piece of steel.
- **Hole saw:** The name of this tool describes exactly what it does. Hole saws are round and come in different sizes to cut holes of a specific diameter. The cheap ones are made of steel. Several sizes of blades fit into a round body assembly called an *arbor* that has a drill in the center. This type of hole saw is capable of sawing through ¾-inch material. To drill through larger material, you need a hole saw with hardened-steel or carbide teeth.
- **Close-quarters drill:** The *chuck,* which holds the bit of a drill, is angled at 55 degrees to give you easy access to hard-to-reach areas, such as between floor joists. It operates and has the same capabilities of a standard ⅜-inch drill.

  When the drill is plugged in, keep your finger off the paddle switch until you're ready to use the drill. The unexpected whirling of a spade bit can do a lot of damage.
- **Seat dresser:** The seat dresser is a tool that's used to remove calcium on valve seats, so that when the valve stem is closed, it stops the water flow (see the "Finding Plumbing Replacement Parts," earlier in this chapter, for more on valve seats and stems). You can only use this tool for faucets that use leather or plastic valve seats.
- **Plumber's die and die stock:** This tool is essential if you need to thread iron pipe. The die stock is a two-ended handle with a circle in the center that holds the *die,* which is a round piece of hardened steel with several half-holes and sharp internal teeth. Used together, these tools cuts threads on the outside of a metal rod.

- **Pipe cutter:** This large, heavy-duty cutter is capable of cutting iron pipe. It works similarly to a tubing cutter, but has long handles for greater leverage.
- **Plastic tubing cutter:** This tool is designed for quick, clean cuts through plastic pipe and tube. Plastic tubing cutters feature compound leverage ratchet mechanisms and hardened steel blades. They make cutting pipe a one-hand operation.

- **Plastic-cutting saw:** This is an inexpensive saw designed for cutting plastic pipe, plywood, and veneers. It has an aluminum or plastic handle and 18-inch blade with replacements available either 12 or 18 inches long.
- **Pipe reamer:** After cutting metal pipe, you'll always find burrs or small ridges called *flanges* on the inside of the pipe. If the burrs or flanges are not removed, water flow is disrupted, which leads to calcium buildup. Some cutters have built-in reamers. Or, you can use a half-round file. But a pipe reamer with a self-feeding spiral design, shown in Figure 3-11, makes the job fast and easy.

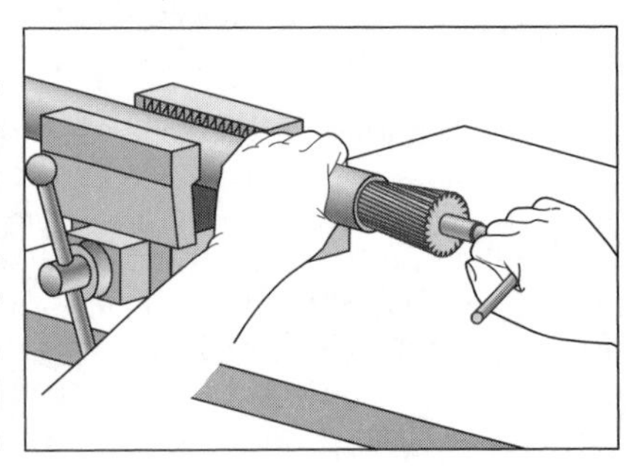

Figure 3-11: You can use a file or sandpaper to remove burrs, but a pipe reamer is helpful if you have several joints to prepare.

- **Inside/outside reamer:** This tool has a plastic housing with steel blades on the inside, which makes it handy for quick, clean, and easy inside reaming and outside beveling on ¼- to 1½-inch plastic, copper, or brass pipe. This tool isn't for use on iron pipe, however.
- **Spring-type tube bender:** This tool is a tightly coiled spring that aids in bending soft copper and aluminum tubing without crimping or flattening it. The benders come in sizes to fit the tubing. To use the tool, slip the bender over the tubing with a twisting motion, and then place it over your knee and bend it slowly.
- **Flaring tool:** These tools are used for flaring soft copper pipe when used with flare nuts. You can find this kind of pipe used on icemakers and humidifirs. Place the flare nut on the pipe, and then clamp the pipe in the proper size hole in the vise bar. Tighten the clamp to close the two halves of the vise bar tightly around the tubing. Slide the c-shaped die clamp over the vise bar and position the cone-shaped die over the mouth of the tubing. Tighten the large screw that the die is attached to, which pushes the die into the tubing and forces the edges of the tubing outward, thus creating the flare.

Plungers, snakes, and augers

The tools in this category are the special-use tools Tim Taylor of *Tool Time* wouldn't be without. They give power to the weak when dislodging clogs in drains, drilling large holes, providing the oomph for bending pipes, and much, much more.

- **Toilet plunger:** Similar to the old-fashioned sink plungers, the rubber cups on these plungers have an extension that fits tightly into the toilet bowl. The extension can be folded back when using the plunger for sinks. For the plunger to work properly, seat the ball in the bottom of the toilet, push down gently, and then pull up quickly.
- **Toilet (closet) auger:** When the plunger doesn't work, the toilet (or closet) auger, shown in Figure 3-12, is the next option. This tool is designed to fit a toilet bowl and clean out the trap. Fit the handle into the bowl, and turn the crank while slowly pushing the flexible shaft through the hollow rod until it hits the blockage. (Chapter 7 has more hints for unclogging a toilet.)
- **Powered closet auger:** Before you call in a plumber, rent this tool, just to make sure that the toilet is really clean. Instead of cranking by hand, a drill-like driver powers the shaft.

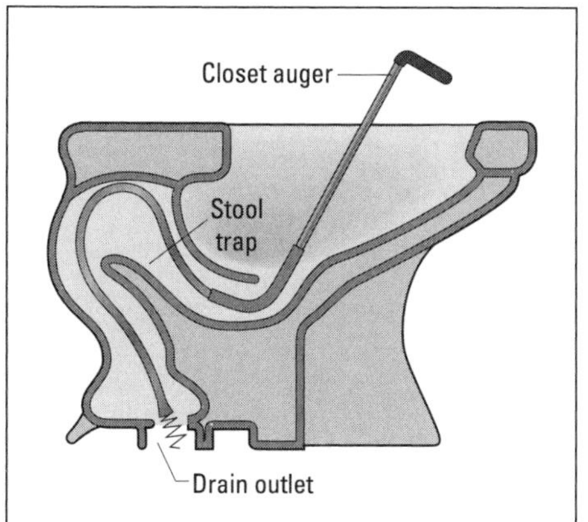

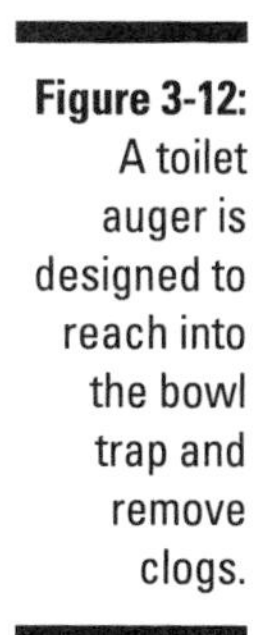

Figure 3-12: A toilet auger is designed to reach into the bowl trap and remove clogs.

Chapter 4

Cutting and Joining Pipe

In This Chapter

- Working with flexible and rigid plastic pipe
- Working with drain-waste-vent pipes
- Using copper piping

In a perfect world, all of the plumbing pipes and lines in your house would be made of the same material, so that installing the lines and connecting them together would be a simple task. Although the job isn't necessarily difficult when your pipes are made of different materials, you do need to know some background information about pipes so that you can tackle these jobs with ease. We cover all of this information in this section, giving you everything you need to know about cutting and joining pipes. And to facilitate joining plastic and metal pipes, we go over the transition fittings and adapters that you need to use.

Working with Rigid Plastic Pipe

A process known as *chemical welding* joins rigid plastic pipes together. You don't use a welding machine for this process; instead, you just use a solvent cement that dissolves the surface of the pipe and the fitting so that they bond together. Each type of pipe — PVC, CPVC, and ABS — requires its own particular formula of solvent cement (see Chapter 3). PVC and CPVC must be cleaned with a primer before you cement them together.

Even if you've joined PVC for non-plumbing purposes without using the primer and the pipe seems to weld together just fine, keep in mind that if you want the pipe to withstand water pressure, you need to clean it with the primer first to ensure a good weld. Having a pipe burst inside a wall is no one's idea of fun, so put the work in beforehand rather than having to deal with a bigger mess later.

The primer and solvent that join plastic pipes together are sold in cans of various sizes. If you have a small job, buy the small cans rather than thinking you'll just have some extra handy in case you need it later. We find that after a can has been opened, the product doesn't keep very well — it soon evaporates or dries into a gooey blob. And that isn't any help to you when you're in the middle of a plumbing project.

When you purchase CPVC pipes and fittings, make sure that the same manufacturer makes the pipe as makes the fittings. If you buy pipes and fitting that haven't been made by the same manufacturer, they could have a poor fit and not create a tight bond when they're welded together.

Cut the pipe to length using a saw and a miter box (see Figure 4-1) or a tubing cutter. If you don't have a miter box to cut the pipe, slide a slip coupling over the end of the pipe. A *slip coupling* has no inside ridge the way that a standard coupling does, and can slide up and down the pipe. Slide it down to the cut. Run your marker around the pipe using the edge of the coupling as a guide. Doing this gives you a straight square line to guide your cut.

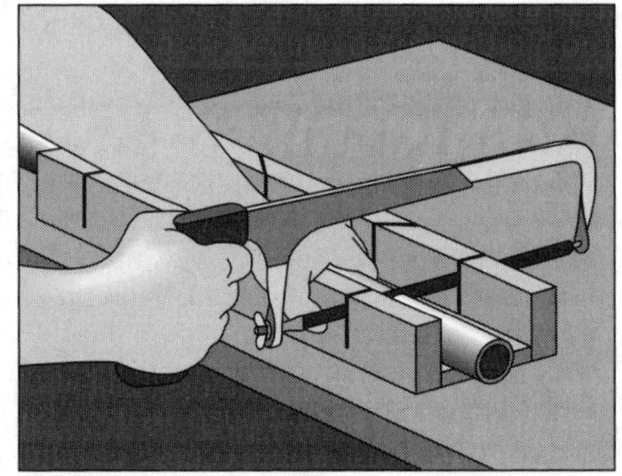

Figure 4-1: You can easily cut plastic pipe with a miter box and a plastic-cutting saw or hacksaw.

Before chemically welding parts together, make a dry run of the pipe and fittings to make sure the pipe is cut to the correct length and that everything fits together as you planned. Mark each fitting and pipe with a pencil line so you can place the fittings in the proper position after you apply the solvent.

Use plenty of ventilation where you're working. The fumes from the primer and solvent aren't good for your lungs. Also, avoid getting the primer and solvent on your hands by wearing chemical-resistant gloves while you work with these chemicals. Finally, keep in mind that you won't be able to remove any cement that drips on your clothes, so wear ones that you don't care about.

If you need to join CPVC or PVC pipes, follow these easy steps:

1. **Smooth the inside of the pipe with a knife, fine (120-grit) sandpaper, or a deburring tool.**

   Bevel the outside edge of the pipe to ensure that the solvent isn't forced out of the fitting.

2. **Clean the joining surfaces on the outside and inside of the pipe. See Figure 4-2.**

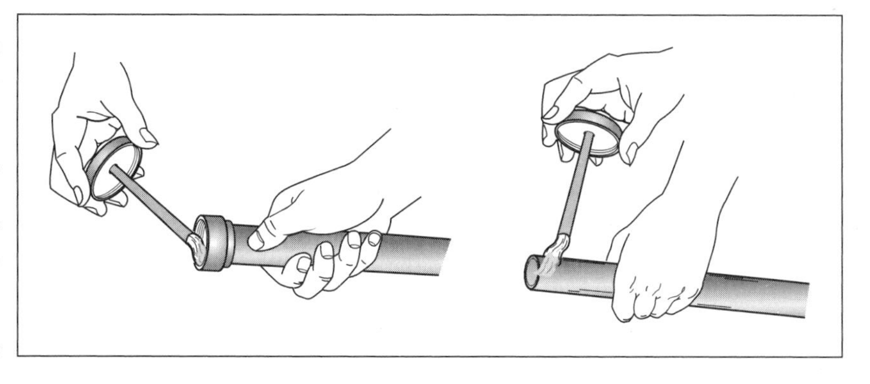

Figure 4-2: Clean the pipe, and then apply a liberal amount of cement to the pipe and fitting.

3. **Apply primer to the inside of the fitting and outside of the pipe.**
4. **Wait about 15 seconds after applying primer, and then apply a liberal amount of solvent cement to the pipe and fitting, as shown in Figure 4-2.**
5. **Push the pipe into the fitting about ½ inch off the marked lines.**

   Be sure that the pipe is fully seated, twist the fitting or pipe to spread the cement, and then align the marks. Take a look at Figure 4-3 for an example.

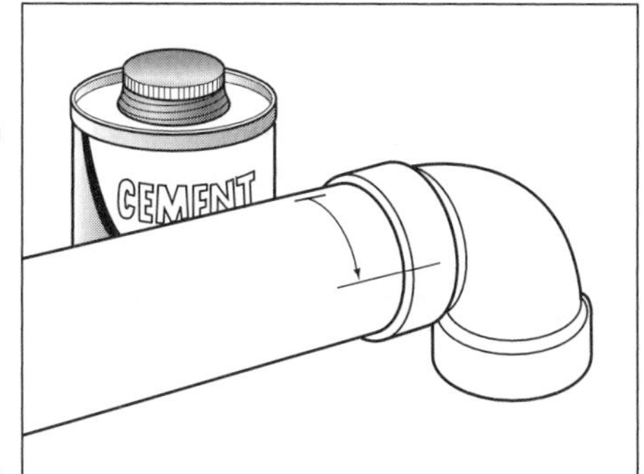

Figure 4-3: Twist the fitting on the pipe and align the marks quickly.

Work quickly because the cement sets quickly — in less than 60 seconds for ½-inch pipe and within a minute or two for 1½- or 2-inch pipe.

6. **While you're moving the fitting, be sure that you've put a continuous bead of cement around the joint.**

   If the cement is not continuous, quickly pull it apart and apply more cement. If you make a mistake and the pieces are glued together, you're better off cutting out the fitting and starting over.

Solvent cement sets quickly and although you can salvage a fitting, doing so is usually not worth the effort. If you make a mistake or if the joint leaks, cut out the offending part and add a coupling. Use the same gluing methods just described.

Joining plastic pipe to a water heater

Use only CPVC plastic pipe when you're joining a pipe to a water heater. CPVC is rated to withstand pressures up to 100 psi and temperatures up to 180 degrees Fahrenheit, but placing it too close to a water heater isn't a good idea. Instead, install a 12-inch-long galvanized steel nipple between the heater and the plastic pipe. Then, to connect plastic pipe to a metal pipe, follow the instructions in the following section. (Chapter 18 has more information on replacing a hot water heater.)

Joining plastic pipe to metal pipe

To join rigid plastic pipe to other kinds of pipe, you use *transition fittings* or *adapters* that can be attached to metal pipe. For connecting copper pipe, you use two threaded fittings. One is solvent welded to the plastic pipe and the other fitting is soldered to the copper pipe. Because of the heat involved, you should solder the copper fitting first (before attaching the plastic pipe). Then attach the plastic fitting to the copper fitting. Finally, weld the plastic fitting to the plastic pipe. See Figure 4-4.

If you're not able to join the pipes in this manner, or if you want to separate them in the future, use a *union.* Unions are available for both copper and plastic. Copper unions are more durable, so you're better off installing a union on the end of the copper pipe, and then using a copper-to-plastic adapter to join the two types of pipe together.

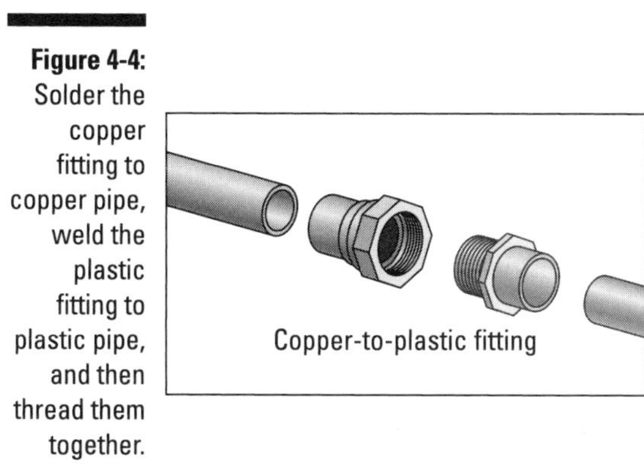

Figure 4-4: Solder the copper fitting to copper pipe, weld the plastic fitting to plastic pipe, and then thread them together.

Using Flexible Plastic Pipe

Polybutylene (PB) piping is a gray-colored, flexible, plastic pipe. It's easy to install and you can snake it or fish it through hard-to-access places, which is a great advantage for renovation or remodeling work. Today, the use of PB piping for home water supply pipes is against code in many localities, mostly because the *acetal* (plastic) fittings used in PB piping throughout the 1980s developed leaks. The pipes leaked primarily because the crimp rings used to tighten the pipe to the fitting tended to crack the fitting, making them more susceptible to chlorine degradation and, subsequently, leaks.

Another product, *cross-linked polyethylene* (PEX), is being used as an alternative to PB in many parts of the United States. While PEX is relatively new in the United States, it's been used in Europe for more than 25 years. Its first use in the U.S. was for radiant floor-heating systems. Check with your building code officials to see if it can be used where you live — this type pipe is now being used throughout the U.S.

Joining flexible plastic pipe

A variety of compression fittings are available for joining flexible plastic pipe. Some can be used to connect plastic pipe to copper or iron pipe. Because of past failures of these products, however, some manufacturers warn against using them in behind-the-wall installations. You can interpret the warning to mean "It may fail" or "We don't want to be sued if it does." But these fittings are still a good choice for making temporary repairs, or for use on exposed pipe, like you have in your basement.

You can join flexible plastic pipe in one of several ways. For example, in cold-water applications like a lawn sprinkler system, you can use simple fittings and hose clamps to joint the pipe. This type of fitting has a male end, which has a ridged surface that's designed to be pushed into the end of the pipe. The ridges act like barbs to prevent the fitting from coming out of the pipe after the hose clamp is tightened.

With in-house plumbing, however, compression fittings are required.

Compression fittings are available in valves, unions, and couplings. You assemble all of them in the same way:

1. **Cut the end of the tubing square.**
2. **Slide the compression nut over the tubing.**
3. **Push the compression ring, flange, and o-ring onto the tubing.**
4. **Assemble the fitting by carefully threading the compression nut onto the fitting body.**
5. **Tighten the compression nut.**

   You may have to use two wrenches to do this. Follow the manufacturer's directions to figure out how tight to make the joint. Over-tightening can cause a leak.

Fixing leaks

What do you do if you have PB piping installed throughout your house and it develops a leak? First, turn off the water supply and isolate the leak, if possible. Then call a plumber. Plumbing professionals suggest replacing acetal (plastic) fittings with brass or copper fittings, using copper crimp rings around the pipe to seal the joints. When properly installed, these fittings eliminate the leaking problems.

If the pipe is behind a wall, you have to remove and replace drywall, and repair any damage caused by the leak. This is work you can do yourself — it's covered in our book, *Home Improvement For Dummies* (IDG Books Worldwide, Inc.). Repairing the pipe, however, is not a do-it-yourself project. You need special scissors (which require constant sharpening) to cut the pipe. You also need a special crimp tool to install the rings. So you're better off leaving these repairs for a professional plumber. After this installation is complete, it must be checked with a *go/no go gauge,* to be sure that the ring is properly crimped.

Using poly risers

Poly risers are flexible tubes that are used to join water supply pipes to faucets and toilet tanks. They're made of CPVC or PB and are inexpensive. Because over-tightening and fatigue can lead to failure (and leaks) in the standard models, we recommend using *braided poly or nylon risers.* The braiding protects and reinforces the plastic material and minimizes failures.

Two types of poly riser tubes are available.

- One type is designed to be installed between the shutoff valve and a sink or toilet. This riser has a rounded, cone-shaped nose at the top of the tube, which fits into the end of the pipe coming out of the bottom of the valve assembly.
- The second type is designed to connect a toilet with its shutoff valve. This riser has a flat-top end.

Working with Drain-Waste-Vent Pipes

Working with plastic drain-waste-vent (DWV) pipes is essentially the same as working with flexible plastic pipes except that the DWV pipe is larger. Today, PVC or ABS is usually used for drain-waste-vent pipes — you may still find cast iron in older homes, but fortunately, plastic is almost a universal replacement.

The basic difference between water-supply fittings and those designed for drainage is that water running through drainage fittings is only powered by gravity. To promote the smooth flow of water, the fittings are designed so that the wall of the pipe is flush with the inside diameter of the fitting. Bends in these fittings are more gradual and have a larger radius than pressure fittings, because drainage fittings are designed to carry water that's not under pressure.

Working with Copper Pipe

For plumbing work, copper pipe has many advantages: It's universally accepted by plumbing codes, it's long lasting, it resists mineral deposits, and it can be used with either mechanical or soldered connections. After you know how to solder, copper pipe is easy to work with, modify, or repair. But compared to plastic, it's expensive, which is why copper is seldom used for residential drainage and vent systems — plastic rules the day for these applications. For water supply piping, however, copper is the material of choice.

Rigid copper makes a neat appearance in long runs, because the rigid pipe can easily be supported by the house structure, but it requires you to use more fittings. The right-angle fittings that are typically used to change direction offer more resistance to water flow, making your system less efficient.

Cutting copper pipe

You can cut copper pipe with a hacksaw, but using a tubing cutter (covered in Chapter 3) is a better choice. Follow these steps:

1. **Make a mark on the pipe where you want to cut it.**
2. **If you're right-handed, hold the longer part of the pipe in your left hand and the tubing cutter in your right hand.**

   Reverse this process if you're left-handed.
3. **Place the tubing cutter on the pipe with the tubing cutter wheel on the marked cut line. Tighten the thumbscrew so that the cutter blade just barely cuts into the surface.**
4. **Rotate the tubing cutter counterclockwise for one turn to help the tubing cutter wheel make one straight cut around the pipe.**
5. **Make the rest of the cuts in a clockwise direction, tightening the thumbscrew slightly during each turn.**

   Keep turning and tightening until the cut is complete. Don't overtighten the thumbscrew, because you may crush the pipe. After the pipe is cut, use a reamer (covered in Chapter 3) to remove burrs from inside the pipe.

Soldering copper pipe

A lot of people tell us that they can't solder copper pipe, but it isn't difficult if you use the right materials and properly prepare the pipe. You need the following materials, all of which are described in Chapter 3:

- Flux: Choose a self-cleaning, non-acid flux.
- A small brush
- Solder: Don't buy *acid core solder;* it doesn't work in plumbing.
- Propane torch or Mapp torch. Propane torches are the least expensive and the propane gas cylinders are readily available. Mapp heats to a higher temperature, but the torch and Mapp gas cylinders are more expensive.

Don't use Mapp gas with a propane torch. Mapp gas burns hotter than propane and isn't necessary for soldering plumbing fittings.

- Rags

Most building codes require the use of leadless solder. This type solder does not contain lead, so it requires less heat than the 50/50 or 60/40 (lead-tin mix) solders of yesteryear.

In theory, if you use a self-cleaning flux and if the pipe doesn't have dirt or green corrosion on it, you don't have to clean the joints. To make sure that the pipe is clean, though, use a brush or other pipe-cleaning tool to clean the inside and outside of ½- and ¾-inch pipe. You can also use fine (150-grit) emery cloth or steel wool.

If you're installing a new run of pipe, you can add several fittings and fasten the pipe to studs or joists before you begin soldering. Leave a male end to which you can add the next run of pipe. Just remember to solder all joints on a fitting before you turn on the water!

Soldering is easy but take a few precautions: Before you begin any soldering project put on your safety glasses and a pair of work gloves. Keep in mind that you're working with an open flame, so be careful not to burn the house down!

Here's how to solder copper pipe like a pro.

1. **Apply flux to the end of the pipe (see Figure 4-5) and slide on the fitting. Wipe off any visible flux.**
2. **Unwind several feet of solder from the spool and wrap it around your hand.**

   This is easier to handle than a heavy spool.
3. **Make a 90-degree bend in the end of the solder.**
4. **Light the torch and adjust the flame.**

   It's dangerous to use a torch around wood. Always place a piece of metal between the fitting and any nearby wood. A double thickness of heat-vent sheet metal works well, but don't use an aluminum sheet. Keep a fire extinguisher close at hand.
5. **Begin heating the entire fitting. Apply heat to the fitting, rather than to the pipe. Wipe away any flux that runs down the pipe.**

   See Figure 4-6. Move the heat around the fitting for a few seconds. With the torch to one side, apply solder to the joint on the opposite side. When the solder begins to stick, you're close to being done.

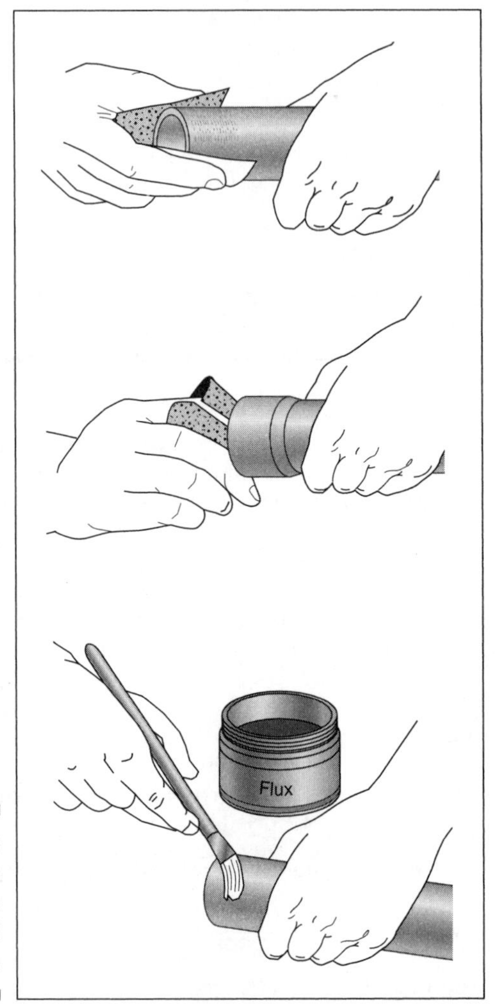

Figure 4-5: Apply flux to end of pipe with small brush.

6. **When the solder begins to flow into the joint, remove the torch and continue to feed solder until it drips out of the bottom of the joint.**

   You need about ½ inch of solder for a ½-inch pipe joint; a ¾-inch joint takes about ¾ inch of solder. Work from the bottom of each joint and move up.

7. **Wipe away excess solder.**

A properly soldered joint draws solder into the joint and has no pinholes. If this isn't the case with your newly soldered joint, reheat the joint and add more solder. If the problem persists, the joint wasn't clean. You have to remove the fitting, clean it again, and resolder.

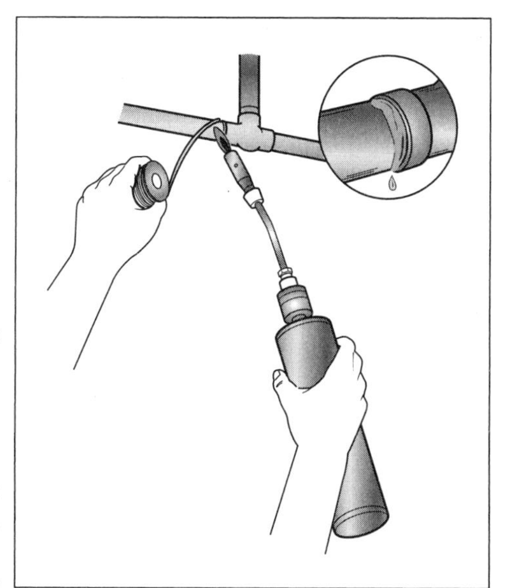

Figure 4-6: Solder all joints on one pass, beginning at the lowest point.

Soldering brass valves

There are two schools of thought regarding soldering valves. One is that you remove the stem to protect the valve seat (see Chapter 3 for a description of valves and their parts). The other is that removing the stem isn't necessary — just open the valve and direct the flame away from the valve body (see Figure 4-7) — because some valve stems are difficult to reseal after they've been removed. Of course, if you melt the seal, that's a moot point. We suggest that you remove the stem if you're not experienced with soldering, and open the valve if you are.

When soldering copper to brass, remember that the brass is denser and requires more heat. Direct the flame to the brass and move it around the fitting to get better distribution of the heat. Be careful not to overheat the copper. If copper begins is change color, it's too hot — you'll burn the solder and the joint will leak.

Using a torch around wood is dangerous. Always place a piece of metal between the fitting and any nearby wood, and keep a fire extinguisher close at hand.

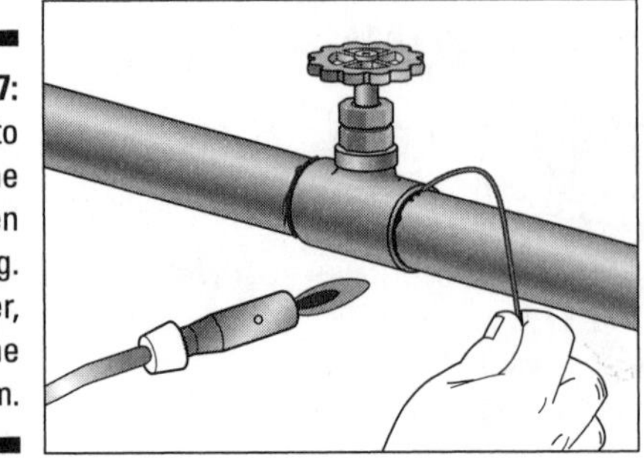

Figure 4-7: Be sure to open the valve when soldering. To be safer, remove the valve stem.

Separating solder joints

When you need to make changes to your water supply system, whether it's to fix a leak or because you need a new bathroom, you may have to separate a soldered joint that has served you well for many years.

To successfully unsolder a joint in a water-supply pipe, you first have to make sure that you don't have any water in the pipe. If you heat the pipe while there is water in the pipe, the water turns to steam and cools the pipe, making soldering impossible. Try these ways of getting water out of your pipe:

- Turn off the water supply valve (see Chapter 2) and then open the faucets throughout the house.
- Open a faucet at the lowest point in the water system — it may be an outdoor faucet or a basement toilet — so that the water has an escape route.
- If turning off the water supply and opening a low-point faucet doesn't work, cut the pipe at a low point to let the water out (see the "Cutting copper pipe" section, earlier in this chapter). To repair a pipe that's been cut at a low point, solder in a union, and then thread the union together and tighten. The union will be easy to open again, if needed.

After all of the water is removed, heat the joint with a propane torch. You can tell when the pipe is heated enough by watching the solder around the joint — it changes slightly from a dull silver to a more shiny look as it melts. Using groove-joint pliers (see Chapter 3), grip the fitting and apply a twisting motion it until the fitting begins to move. Continue to apply heat and slip the fitting off the pipe.

Don't touch the pipe or fitting with your bare hands; the pipes and fittings are hot!

Working with Flexible Copper Pipe

For residential use, copper pipe comes in two varieties: hard (also called rigid) and soft. Soft copper will often take a freeze without springing a leak, but rigid copper seldom will. Soft copper pipe is a good choice in remodeling situations where pipe has to be snaked around obstructions or through small openings.

Rigid copper pipe is available in 10- and 20-foot lengths, while soft copper comes in 60- and 100-foot coils. If you need a shorter length of pipe the hardware store will usually cut it for you. Soft copper is typically joined with flared or compression fittings.

You can find a whole range of fittings made to join copper tubing. These fittings are available in either compression or flared-type. The compression fitting uses a small copper ring that's squeezed tight around the tube by to form a tight seal. A flared-type fitting is designed to tightly grip the widened (or flared) end of the tube after it's prepared with a separate flaring tool.

Here's how to install a flared-type fitting.

1. **Unscrew the compression nut from the fitting and slide it on the tube.**

   Make sure the threaded end faces the end of the tube so that you can screw it on the fitting.

2. **Use a flaring tool (see Chapter 3) to flare the end of the tube.**
3. **Apply a bit of pipe joint compound (see Chapter 3) to the flared end of the tube, and then push the flared end of the tubing against the fitting.**

Working on gas piping

Before you begin working on the black iron pipe that brings natural or liquid propane (LP) gas into your home, check with your local code officials to see if you're permitted to work on this pipe — because a gas leak can be extremely dangerous, many codes prohibit the practice. Gas piping repairs have led to house explosions that have been disastrous. If your area doesn't allow it, hire a plumber.

If you are allowed to work on your gas piping and you decide to do so, have the gas company, a codes official, or a professional plumber check your work. This is not a project to treat lightly — be extremely careful.

Keep in mind that you may also have to make several changes in your system to bring it up to code, which in many areas has recently changed. One person we know had to remove the aluminum tubing that serviced a gas clothes dryer. He also had to remove the copper tubing that serviced an outdoor gas grill, and the pipe and fittings to the gas fireplace had to be black iron.

4. **Tighten the compression nut by gripping the nut with one wrench and the body of the fitting with another wrench.**

   Turn the nut clockwise while you use the other wrench to keep the fitting from turning.

To use a compression fitting to join a riser tube to a supply pipe valve, or to join tubing sections together do the following:

1. **Slide the nut and compression ring on the tube (see Figure 4-8).**
2. **Coat both the nut and the compression ring with pipe joint compound.**
3. **Tighten the nut by hand.**
4. **Tighten the nut with two to three turns with a 6-inch wrench.**

   If the joint leaks, tighten only until the leak stops. Overtightening, however, guarantees a leak.

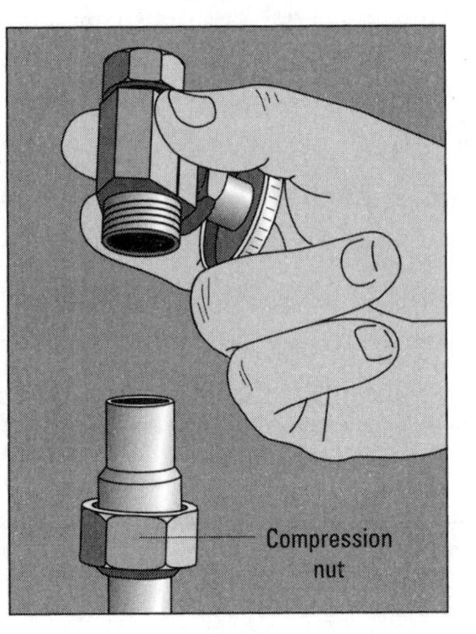

Figure 4-8: Slip the compression nut and compression ring over the end of the tubing, and then insert into fitting and tighten.

Part II
Opening Clogged Drains with Friendly Persuasion

"Be patient everyone. This will just take a minute."

In this part . . .

For clogged drains, Murphy's Law reigns supreme. Rarely does the bathroom sink stop flowing or the toilet get clogged on an uneventful day in the life of your family. It always seems to happen at a time when the last thing you need is another complication.

In this part, we help you through some of life's little traumas so you know how to quickly free a clogged drain, clear a faulty toilet, or retrieve an irreplaceable family heirloom that mysteriously got flushed down a drain.

Chapter 5

Unclogging a Kitchen Sink

In This Chapter

- Unclogging the kitchen sink and faucet aerator
- Clearing stuff out of the garbage disposer

Knowing how to unclog a sink drain is one of those life skills that no one teaches you in any school or university. If you're lucky, when you're young and formidable, you watched your parent or a plumber perform this unpleasant task and paid close attention. If not, well, start reading! This chapter contains the best advice that we can give to help you dislodge the most stubborn clog and get your sink drain back flowing.

This chapter also includes some tips to respond to the cry: "Help, my ring just went down the drain!" or any other drain dilemmas.

Unclogging a Sink Drain

With all the different kinds of food scraps and grease that find their way down the drain, it's not surprising when the drain gets stopped up. Symptoms of a clog can range from water at the sink draining slower and slower to a stagnant pond in your sink.

The easiest solution for drain clogs is to use any of a wide range of chemical drain uncloggers, available in solid and liquid form at supermarkets, hardware stores, and plumbing-supply dealers. You pour the product in, wait for it to dissolve the blockage, and then flush the drain with running water.

Be aware that some chemicals can damage the plastic or rubber parts of a garbage disposer and cause injury if the cleaner splashes into your eyes or onto your skin. If you decide to use chemicals, read the package directions and precautions carefully and follow them precisely; the directions vary by product. If the blockage doesn't clear after a couple of tries, you're ready for a more hands-on approach, covered in the three following sections.

Pouring a kettle full of hot boiling water down the drain is a non-toxic alternative that often eliminates the clog.

Removing the trap

To unclog your sink drain, remove and clean the *trap* (the U-shaped pipe located under the sink), shown in Figure 5-1. Removing and cleaning the trap is easy.

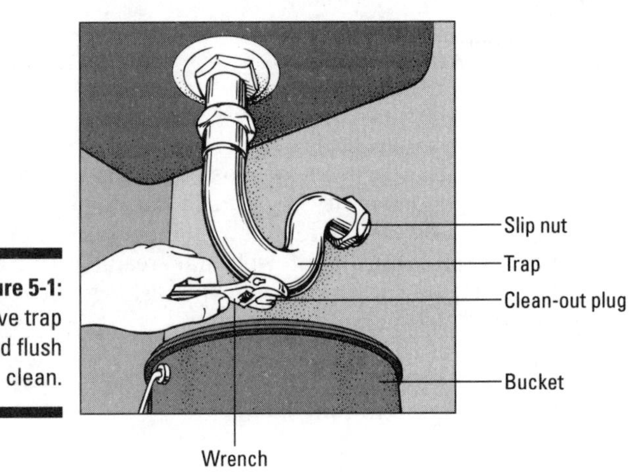

Figure 5-1: Remove trap and flush clean.

1. **Place a bucket under the trap (before taking it apart).**

   This way, you catch any debris or water that falls out when the trap is removed.

   Before removing the trap, wear rubber gloves and eye protection from any sharp pieces of debris or whatever has caused the blockage.

2. **Use a wrench or slip joint pliers (described in Chapter 3) to unscrew the metal slip nuts about a half turn or so, so that you can loosen them by hand.**

   Some traps have a clean-out plug instead of slip nuts. Simply remove the plug and allow the blockage to spill out.

   To protect the chrome finish on the slip nuts, wrap tape around the jaws of your wrench or pliers. Plastic traps have slip nuts that you can usually turn by hand.

3. **Scrape out any blockage from the trap.**

Found objects

It has happened to everyone: A treasured ring or favorite toy goes sailing down the drain, and you're desperate to retrieve it. The best rescue tactic is to remove the trap, hoping that the object is heavy enough to settle in the lower part of the trap. Here's what you do:

1. **Don't run any water through the drain which may flush it further away.**
2. **Get a bucket and a wrench, and then follow the directions in the "Removing the trap" section of this chapter.**

   Keep the bucket under the trap to catch what's inside it when it's removed from the drain pipe.
3. **Good luck!**

4. **Tighten the slip nuts with your hands to be sure that they are threaded on the trap correctly, and then tighten with a wrench or pliers.**

   A half a turn is usually all that's necessary to stop the trap from leaking; don't overtighten.

Some folks may advise you to try unclogging the sink drain with a plunger (covered in the following section) before you resort to removing the trap. We believe that cleaning the trap first is a better approach, because using a plunger can push the clogged material from the trap into the drainpipe, where it's more difficult to remove.

A homeowner's best friend — the plunger

A common plunger is capable of unclogging a drain that even the toughest chemicals cannot budge. Unlike chemicals, a plunger uses suction to alternately push and pull the clog within the pipe until the force dislodges the blockage. If cleaning the trap doesn't clear the clog, try plunging the offending clog.

Don't confuse a common plunger that's used for drains with a *toilet plunger,* which has two cups, one inside the other. A *common plunger* has a wooden broomstick-like handle that attaches to a cup-shaped piece of rubber.

Before you start whaling away with the plunger, remove any standing water that may contain chemicals. Splashing diluted chemicals into your eyes can cause severe damage.

Here's the plunger procedure:

1. **If the sink has a stopper, try to remove the stopper first, in order to give you a wider opening to the drain.**
2. **Pour a full kettle's worth of boiling water down the drain to break up the clog.**
3. **Fill the sink with enough tap water to cover the rubber portion of a plunger, thus assuring good suction.**
4. **Place the plunger over the drain and vigorously push down and pull up several times.**

   If you're successful, you'll notice a sudden emptying of the sink.

The snake

Suppose neither cleaning the trap nor plunging clears the clog. Your final weapon is a *drain auger,* or *snake.* This tool, a coiled spiral snake that's usually about ¼-inch thick with a handle on one end, works the opposite way than a plunger does: You push the snake into the clog in the water and crank it to drive the snake further into the obstruction, as shown in Figure 5-2. While parts of the clog break up and flush through the drain, the snake helps you gain access to the clog so that you can pull it out. Some snakes can fit as an attachment on an electric drill, giving it more power to force it through the clog. Snakes are especially handy because they're long enough to reach clogs that are deep in the drain pipe.

An ounce of prevention . . .

We may be too late with this information, but your best defense against clogs is to avoid them in the first place. The following are some common-sense practices to use:

- **Use a sink strainer:** A clogged kitchen sink is usually the result of garbage or foreign objects entering the drain. Use the sink strainer to prevent garbage and small items from entering the drain pipe.
- **Take care of your garbage disposer:** When using a garbage disposer, run cold water at full volume while the machine is chopping the garbage; leave the water running for a full minute after you shut the disposer off. This precaution flushes the garbage completely out of the small-diameter sink drain pipe and into the larger main drain pipe, where it's less likely to cause a clog.
- **Don't dump materials down your drain:** Building materials flushed down the drain by do-it-yourselfers (the most common offender is plaster or wallboard compound, which seems innocent enough going down, but can harden in the drain pipes and clog them). To prevent these clogs, never dispose of leftover building materials in sink drains.

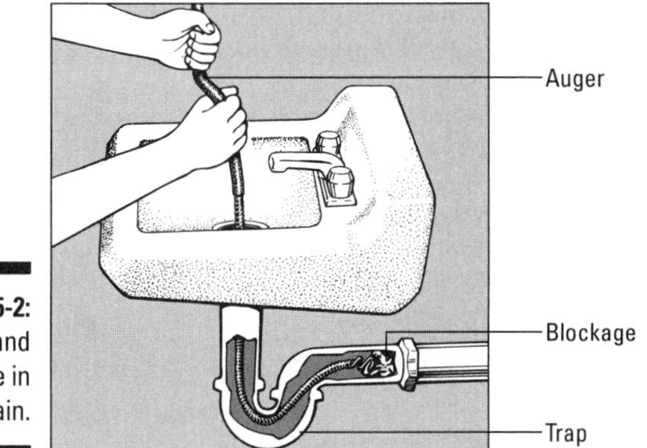

Figure 5-2: Snake and cable in drain.

You can rent a manually operated or electrical drain auger for a few bucks at a rental center. The equipment is easy to use, but ask your dealer for instructions about how to operate it.

The basic process is as follows:

1. **Push the end of the snake into the drain opening and turn the handle on the drum that contains the coiled-up snake.**

   The auger then begins its journey down the drain.

2. **Keep pushing more snake into the drain until you feel resistance.**

   You may have to apply pressure when cranking the handle to get it to bend around the tight curve in the trap under the sink. After turning the curve, the snake usually slides through easily until you hit the clog.

3. **Rotate the snake against the blockage until you feel it feed freely into the pipe.**

   The rotating action enables the tip of the snake to attach to the clog and spin it away or chop it up. If the clog is a solid object, the auger head entangles the object. If you don't feel the auger breakthrough and twisting getting easier, pull the auger out of the drain — you'll most likely pull the clog out with it.

4. **Run water full force for a few minutes to be sure that the drain is unclogged.**

   Sometimes, the clog flushes down the drain; other times, the clog comes out attached to the snake.

If the snake doesn't fit down the drain or gets held up in the trap, you need to open the trap beneath the sink. Follow the instructions in the "Removing the trap" section, earlier in this chapter. Avoid contact with the water that comes out from the trap, because it may contain chemical drain opener. From the trap, insert the snake in either direction until you reach and clean out the clog.

Unclogging a Faucet Aerator

If a faucet seems to be running slower than usual, the aerator may be clogged with a build-up of mineral deposits. An *aerator* is a simple insert that fits inside of a faucet spout's chrome cap — most faucets come with them — to conserve water and keep it from splashing all over the place, while still providing a steady stream and good water pressure. Tiny holes in the aerator restrict water flow by mixing air bubbles into the water stream — see Figure 5-3. The minuscule holes eventually become blocked by small particles in the water; even the ones that get through may be trapped by the aerator screen.

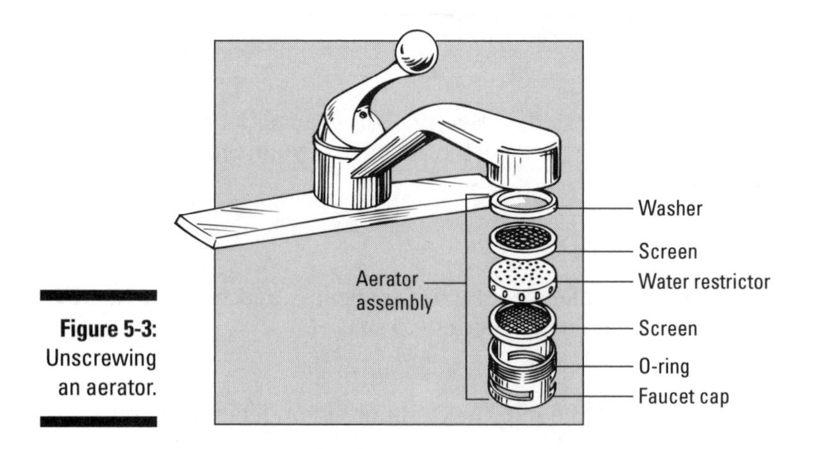

Figure 5-3: Unscrewing an aerator.

Follow these steps (referring to Figure 5-3) to clean and unclog an aerator:

1. **Place a towel or rag over the faucet cap, or cover it with a bit of masking tape.**

   This protective barrier keeps the surface from being marred when you strong-arm the cap off.

2. **Using a wrench or pliers, turn the cap in a counterclockwise motion until it separates from the faucet.**

3. **When the cap is off, remove the screen and water restrictor.**

   Pay attention to the way these small internal parts are arranged. When the time comes to put the aerator back in place, you have to replace these parts in the same sequence and position.

4. **Clean the screens by flushing it with water or using a brush. Push through the tiny holes with a needle or pin to unclog them.**

   Soak an aerator in a cup of vinegar overnight to clean out the small screens. Flush it with clear water before reinstalling it.

5. **Reassemble the aerator in the reverse order that you took it apart.**

You should notice a big difference in the water flow after cleaning the aerator — no, not gushing water, but a nice steady stream.

Unclogging a Garbage Disposer

Even a garbage disposer can be finicky, so don't expect it to devour and digest everything. For example, don't throw corncobs, artichokes, avocado pits, or fish and chicken bones down a disposer and expect it to continue working without a clog. Flip through the owner's manual to find out just what your disposal's limitations are.

Clogs can and do occur, however, and are generally caused by the following:

- Dropping foreign objects — usually a spoon or fork — into the disposal.
- Feeding garbage in too rapidly.
- Failing to run enough water (to completely flush out the drain pipes) while the garbage is being processed.

WARNING!

Never use chemical drain cleaners in a disposer, because the chemicals are highly corrosive, and may damage rubber or plastic parts.

Use Mother Nature's deodorizer for your disposer: Every few months cut a lemon in half, throw it in the disposer, and then turn on the unit. Let it run for a minute or two. The lemon removes the build up of residue on the interior of the disposer and deodorizes the unit. You know it's working by the fresh "lemony" smell.

To unclog a disposer, follow these steps:

1. **Shut off the electrical power switch.**

   This switch is located under the cabinet, near the disposer, or on a wall nearby. If you don't find a switch, go to the main power panel and turn off the breaker or remove the fuse that powers the disposal.

   Never put your hand into the disposer. Remember that the switch may be defective, so keep your hands out of the disposer even when power to the machine is turned off.

2. **Take a look in the disposer.**

   A flashlight may shed some light on the problem — you may see a large object caught in the disposer.

3. **If the stoppage was caused by an object, use a pair of pliers to reach into the disposer and remove it.**

4. **Wait 15 minutes for the disposer motor to cool.**

5. **Turn on the power and push the reset or overload protector button.**

   This button is located on the bottom side of the disposer.

If the disposer is still clogged, follow these steps:

1. **Turn off the power and insert a long dowel, wooden spoon, or broom handle — never your hand — into the drain opening.**
2. **Push the bottom end of the wooden probe against the *impeller* (the blades that grind up the garbage), as shown in Figure 5-4, and rock it back and forth to free it.**

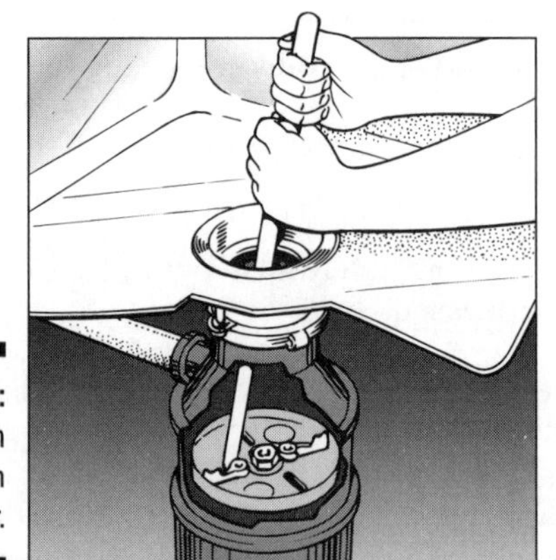

Figure 5-4: Wooden stick in disposer.

3. **When the impeller moves freely, wait 15 minutes for the motor to cool, turn on the power, and push the reset button.**

Some disposer models come with a large L-shaped hex wrench. If you have such a model, turn off the power, insert the hex wrench into the opening in the center of the disposer's bottom, and turn the wrench back and forth until the impeller is freed. Again, wait until the motor has cooled, press the reset button, and then try operating the disposer.

Don't install or use an existing garbage disposal if you have a septic system, which is common in areas where a public sewage system isn't available. A septic system is for waste water and solids, not garbage. Kitchen waste interferes with the normal bacterial action of the system. For the same reason, you should never dispose of chemicals, grease, petroleum products, or anything else that may disrupt the natural bacterial action of your septic system.

Chapter 6

Unclogging a Tub or Shower Drain

In This Chapter

- Keeping your drains running freely
- Finding ways to unclog your shower and tub
- Cleaning and adjusting plunger-type and pop-up drains

Ah, the luxury of taking a relaxing soak in the old bathtub or indulging in a long, hot, steamy shower — everyone has a preference, but when denied, it can be downright annoying. In most busy households, an out-of-commission tub or shower can throw a wrench in kids' and parents' bathroom schedules — a family nuisance to be sure.

A clog in a tub or shower drain usually doesn't come as shock. The blockage comes on gradually. Take action as soon as you see that the tub water doesn't drain as fast as it usually does or when you find yourself standing up to your ankles in shower water that should have gone down the drain. This chapter shows you what to do.

Preventing tub and shower clogs

Tub and shower drains clog up with soap scum and human hair — not a pleasant sight, to be sure. As a defensive measure, fit a strainer over the drain to catch debris before it can enter and clog the drain. Keep the strainer in place and clean it after each use. If you get in the habit to do it after every use (and teach your kids to do the same), the nasty buildup just can't happen.

You can find replacement strainers in the plumbing section of home centers. Measure the diameter of the drain to find one that fits exactly and then simply place it in the drain.

Unclogging a Bathtub or Shower Drain

Sometimes clearing a clogged drain is as simple as pouring a kettle of hot water down the drain to loosen the clog and open the drain. Pour a gallon of hot water down the drain. If it flows down the drain, repeat the process. Voilà — that may be all that you need to do.

Some people rely completely on chemicals to open a clogged drain. It's as simple as pouring a chemical drain opener down the drain, waiting the specified period of time, and flushing the drain thoroughly with running water — be sure to follow product directions carefully, however. You may have to repeat this procedure a few times until the chemicals dissolve the blockage enough to dislodge it. With each application of more chemicals, you may see incremental improvement in the tub's drainage. Chemicals are especially effective for clearing bathtub drains because they contain protein-dissolving elements that can work wonders on the most common cause of bathtub clogs — masses of accumulated hair.

If hot water and chemicals don't work, try using a plunger or snake — both are described in the next two sections.

Taking the plunge

Using a common plunger, also called a *plumber's helper,* is a simple procedure. It has a wooden broomstick-like handle attached to a cup-shaped piece of rubber. This piece of equipment should not be confused with a *toilet plunger,* which has two cups, one inside the other.

1. **Remove the stopper, if there is one.**

   The *stopper* is a pop-up plug that alternately seals or opens the opening or opens the drain — see Figure 6-1. To remove it, push and pull it up and then down until it is released. Raise the *trip lever,* which lifts the stopper up, and remove the entire stopper assembly. If your drain does not have a stopper, remove the drain strainer by unscrewing it or prying it up around its edges.

2. **If you find an accumulation or hair and soap scum, clean the stopper by removing any debris, and then flushing it in a stream of water. Reinstall the stopper assembly in the reverse order that you remove it.**

3. **Stuff a wet washcloth into the overflow holes to plug them up and improve suction.**

4. **Fill the tub with a couple of inches of water, enough to partially submerge the plunger head, to assure airtight suction.**

5. **Push the plunger down and pull it up forcefully.**

   You know that the blockage is dislodged when the water begins to drain much more quickly than it did before.

A bit of petroleum jelly on the lip of the plunger helps form a tight seal, making the plunger more efficient — see Figure 6-2.

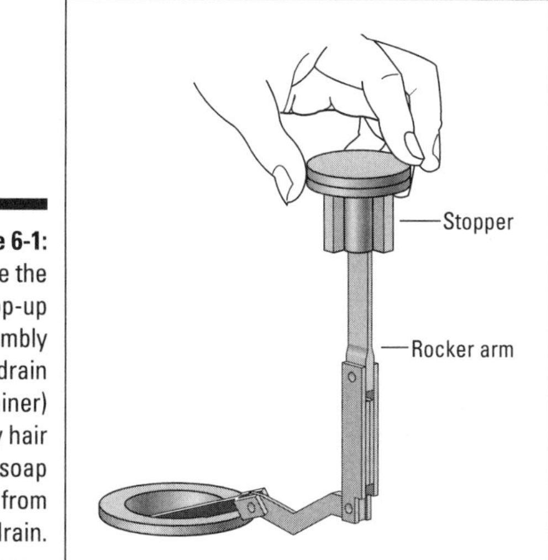

Figure 6-1: Remove the pop-up assembly (or drain strainer) and any hair or soap scum from the drain.

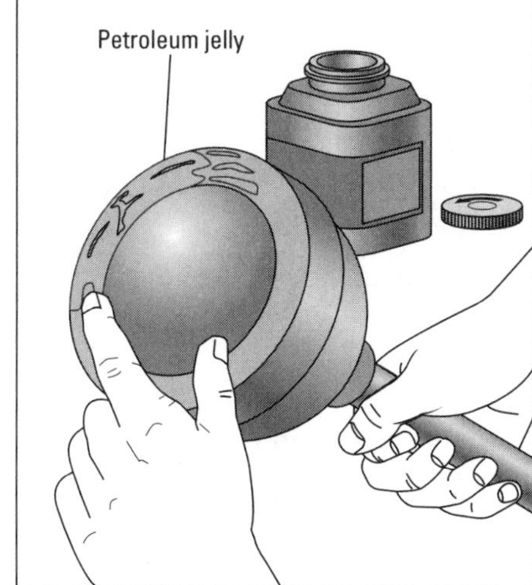

Figure 6-2: Place a bit of petroleum jelly on the rim of the plunger to form a seal between it and the tub.

Charming your tub with a snake

If using a plunger doesn't work, try using a *drain auger,* or *snake,* to unclog the tub drain. For more information on snakes, see Chapter 5.

1. **Remove the tub stopper or strainer, and then push the tip of the snake into the drain opening, continuing to feed the coil through until it meets an obstacle.**

   See Figure 6-3. Push the cable into the clog in the water and crank it further to dislodge the obstruction. Some snakes have an attachment for an electric drill, giving it more power to force it through the clog. To remove a clog that's beyond the trap under the tub, rotate the snake while you push it into the pipe.

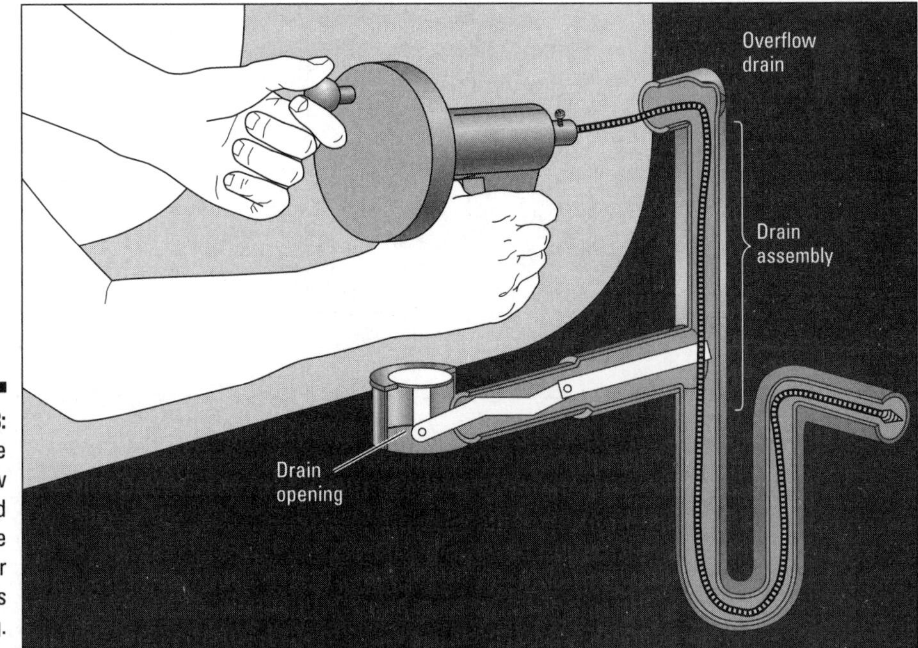

Figure 6-3: Remove the overflow cover and feed the auger through this opening.

2. **If the snake continues to resist passage through the trap, remove the *overflow cover plate* (usually located in the end of the tub directly above the drain and below the faucet).**

   As you remove the screws that hold the cover in place, be careful not to drop the screws down the drain.

3. **If the drain-closing mechanism is attached to the cover, pull up on the cover; the linkage comes out of the overflow hole with the cover.**

4. **Push the snake directly down into the hole, and through the trap.**

   Note that this approach doesn't allow you to get directly to any blockage that's stuck in the trap.

If the blockage is past the trap, somewhere down the drainpipe, reaching it may take some time. The cable stops feeding in when it hits the blockage. You must work the snake against the blockage until the cable freely feeds in the drain — then you know that the clog is clear. Flush the drain with water to confirm that the blockage is gone.

Cleaning and Adjusting Drain Stoppers

Except for really old drains that rely on a rubber stopper to hold back the water, most sinks have a pop-up drain mechanism. Bathtubs have either the same type pop-up stopper or a trip-lever (plunger type) drain closure. When you're faced with a drain that won't hold water, you have to adjust them. The following sections show you how.

Trip-lever drain

A trip-lever drain system has a strainer over the drain opening and an internal plunger mechanism that closes the drain — see Figure 6-4. The plug is operated by a lever located in the overflow plate at the front of the tub. Raise the lever and it lowers a plug into the pipe at the base of the tub, blocking water flow out of the tub. This plug may have a rubber seal on its base that can become old and cracked. Also, debris can get into the seat where the plunger rests causing a slow leak.

Removing the plug, cleaning it, and adjusting the control mechanism cures most problems. Here's how:

1. **Remove the overflow cover plate.**

   The plate is held in place by a couple of screws — remove the screws and pull out the linkage assembly, which is made up if the striker rod, middle link, and plug (see Figure 6-4).

2. **Clean any loose hair or buildup of soap scum from the linkage assembly.**

   You may find many different renditions of this basic design. The drain in your tub may not be exactly like the one illustrated in Figure 6-4, but it'll be similar.

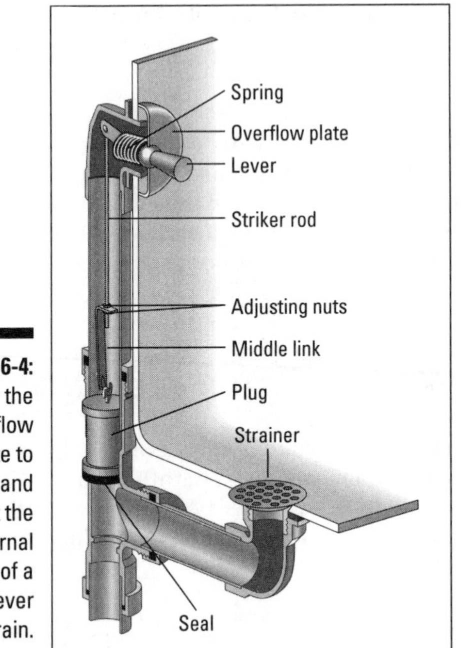

Figure 6-4: Remove the overflow plate to clean and adjust the internal parts of a trip lever drain.

3. **Before you reinstall the plug, inspect the rubber seal (if there is one) at the bottom of the plug.** If it is cracked or broken, take the plunger assembly to a home center or plumbing supply house and get a replacement.
4. **Replace the linkage assembly.**

   You may have to wiggle the plunger a bit to get it to fall back into the drain.
5. **Run some water into the tub.**

   If the tub drains but doesn't hold water, adjust the plug so that it falls deeper into the overflow passage. Remove the assembly, loosen the adjustment nuts and lengthen the linkage controls. A little adjustment — ⅛-inch or so — is all that is needed.
6. **Reassemble the assembly and test. Some readjustment may be necessary.**

Pop-up drain

The pop-up drain assembly, shown in Figure 6-5, has a drain stopper in the opening. A slow-running drain can be the result of the pop-up that's not fully opening. A leaky drain may be the result of a bad rubber seal on the pop-up assembly, or incorrect adjustment of the control mechanism that connects the pop-up to the lever at the end of your tub, which prevents the pop-up from fully closing.

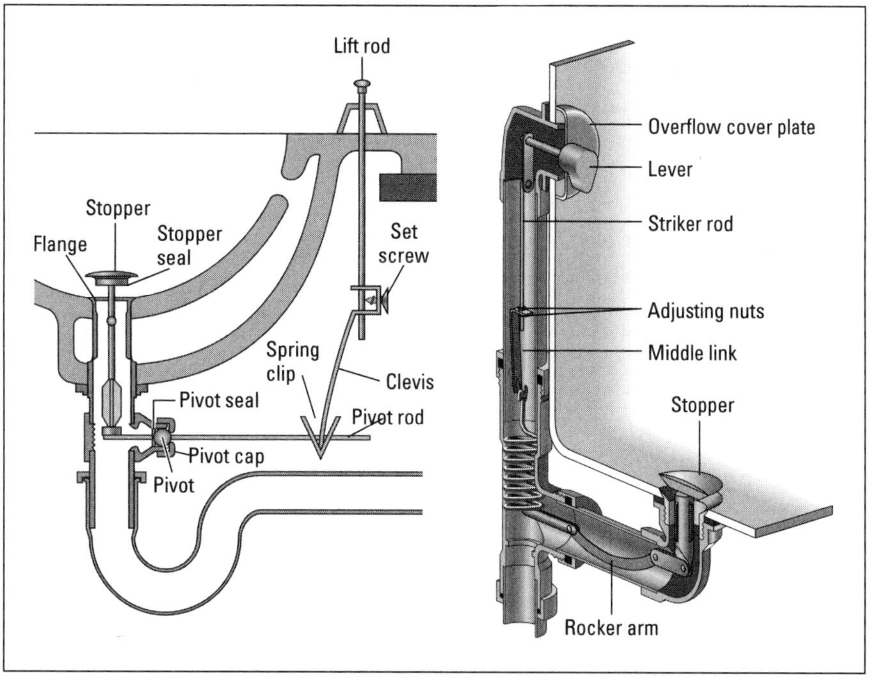

Figure 6-5: Remove the pop-up assembly or drain strainer and any hair or soap scum from the drain.

Removing the pop-up assembly, cleaning it, and adjusting the control mechanism cures most problems. Here's how:

1. **Remove the pop-up drain assembly by pulling it out of the drain.**

   Grasp the stopper and wiggle it around a bit to get it out. The stopper and the rocker arm it is attached to will come completely out of the drain. If there is a clog of hair or debris on the rocker arm, you may have to remove some of this goop before the stopper will come out of the drain.

2. **Remove the tub overflow cover plate by removing the screws and pulling out the assembly.**

   See the "Charming your tub with a snake" section, earlier in this chapter, for more information on the tub overflow cover plate.

3. **Clean the linkage assembly of any loose hair or buildup of soap scum.**

   This assembly is composed of the crank lever in the overflow plate, the striker rod, middle link, and striker spring. The spring at the bottom of the control linkage is a magnet for hair buildup.

4. **Inspect the rubber seal (if there is one) on the pop-up. If it's cracked or broken, take the pop-up assembly to your home center for a replacement.**
5. **Replace the assembly.**

   You may have to wiggle the spring a bit to get to fall back into the drain.
6. **Run some water into the tub.**

   If the tub doesn't hold water, adjust the pop-up so that it completely closes in the drain outlet. Remove the assembly, loosen the adjustment screw, shorten the linkage controls. A little adjustment — ⅛-inch or so — is all that's needed.
7. **Reassemble the assembly and test. Some readjustment may be necessary.**

Chapter 7
Unclogging a Toilet

In This Chapter

- Diagnosing the cause of a toilet clog
- Unclogging your toilet in a variety of ways

While clearing a clogged toilet isn't the most pleasant plumbing chore around the house, no utility is used more often than *la toilette.* Getting it flushing — fast — is a top priority.

Diagnosing the Problem

Sometimes you can see what's clogging the toilet (a wad of paper towels or tissue), but sometimes it remains a mystery. In either case, the chore at hand is getting the toilet flowing freely again, and for that, you need to do a bit of detective work.

When you flush the toilet, which of these results do you see?

- **The water is not swirling down as usual.** Consider yourself lucky; this is a sign of an easy-to-fix problem. The mechanism inside the toilet tank may be clogged or stuck, so the tank isn't filling with water. See Chapter 12 for a quick fix.
- **The water level goes down slowly and only weakly flushes the bowl.** This usually indicates a partial block. Your toilet will probably clog completely the next time it's used, so get out the plunger. Sooner — not later — your toilet will refuse to flush.
- **The water level barely drops (if at all), and then begins to rise past the normal full bowl level.** This action is also an indicator of a partial block — one that may be located beyond the toilet.
- **The water continues to rise past the normal full bowl level until it overflows onto your bathroom floor.** To experience such an event is to understand the true meaning of the word *panic.* Your toilet is clogged. Follow the instructions in this chapter to clear the clog.

Don't flush again if the bowl level rises past the normal height. You may be inviting an overflow.

Clearing a Clogged Toilet

If your toilet plays any of the tricks mentioned in the "Diagnosing the Problem" section you have a clogged or partially clogged toilet bowl. Often, the clog is caused by a blockage in the *trap* — the curved passage inside the toilet bowl.

Don't attempt to unclog the toilet with a chemical drain cleaner. It usually won't work and you end up having to plunge or auger through water that contains a strong chemical. Even if you don't use a chemical drain cleaner, be careful when you handle toilet water and waste — it's laced with bacteria. Thoroughly wash the area, your hands, and your clothing with a disinfectant soap.

Before you attempt to unclog the toilet, clean up any water that spilled out of the toilet onto the floor so that you can work in a clean, dry surrounding.

Partial or total blockage of a toilet requires one of three solutions, covered in the following three sections.

Using a toilet plunger

A *toilet plunger* (a ball or cup-type plunger) is designed specifically for unclogging toilets. The rounded lower surface nests tightly in the bowl, giving the plunger great suction action to dislodge the blockage. With the plunger in place, push down gently, and then pull up quickly to create suction that pulls the blockage back a bit and dislodges it (see Figure 7-1).

Figure 7-1: Positioning the toilet plunger.

Using a toilet plunger (as opposed to a sink-type plunger) keeps you from splashing as much water around because the ball of the plunger covers the entire hole. If all you have on-hand is a small, sink-type plunger (see Chapter 5), try using it — it can't hurt, and it just may work.

Using a toilet auger

If using a toilet plunger doesn't work, try using a toilet auger, which is different from a snake or hand auger (see the following section). A *toilet auger,* also called a *closet auger,* is a short, hollow clean-out rod with a spring coil snake inside that has a hooked end. Attached to the coil is a crank handle that you turn — it's designed specifically to fit into a toilet bowl and clean out a clogged toilet trap — see Figure 7-2.

Follow these instructions:

1. **Pull the spring coil through the hollow handle until about a foot is protruding.**
2. **Insert the auger all the way in the bowl and push the spring coil back through the handle until it rounds the sharp bend in the base of the toilet trap.**
3. **Turn the crank while slowly pushing the flexible coil shaft through the hollow rod until it hits the blockage.**

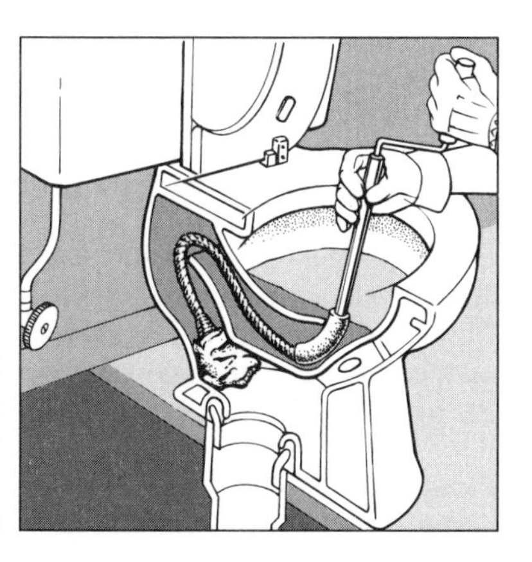

Figure 7-2: Using a toilet auger.

Although the thought of renting or buying a toilet auger may not thrill you, the idea seems amazingly wise when you're faced with a clogged-up toilet. Toilet augers are inexpensive, and they make a particularly handy investment because most toilet clogs occur in the trap — exactly where this tool delivers its punch. Buy one *before* you experience a problem and have to explain to your dinner guests that the bathroom's temporarily closed for repairs.

If you can't get the toilet running freely with a toilet auger, don't flush the toilet, even if it seems to run a little. The water may be backing up in the *soil pipe,* which is the large diameter pipe leading to the sewer or septic system (see Chapter 8). If the water backs up there, it can start running out the basement sink or out of some of the first floor fixtures. Not a pretty picture.

Using a snake

If the toilet auger doesn't do the trick, you probably have a blockage somewhere farther down the line — past the toilet trap and beyond the reach of the toilet auger's coil. Time to rent a snake (see Chapter 5 for details on a snake, or hand auger).

Rent a snake that has a flexible shaft to bend past the tight curve in the toilet trap. The only thing worse than having a clogged toilet is having a clogged toilet with an auger stuck in the bowl; it's tough to explain to the guys at the rental shop why there's a toilet stuck on the end of their snake.

After getting the snake home from the rental shop, follow these steps:

1. **Feed the flexible snake into the toilet until you feel it engage the clog.**

   You know you've hit the clog when the cable becomes harder to turn or refuses to move another inch into the toilet.

2. **Pull the snake back a bit to dislodge the clog.**

   The water level should go down, signaling that the clog is loose.

3. **Flush the toilet to push the clog down the drain line and hopefully, out to the sewer or septic system.**

   If the clog is a diaper or rubber ducky, you may have to pull the offending item all the way out of the toilet to clear the line.

Think of the snake as a powerful tool — use it cautiously so that you don't damage the toilet or drain line with too much force. If the coil becomes really hard to turn, back off a bit and pull it out of the toilet a few inches before attacking again. If you turn the auger too hard, you can kink the wire coil.

Singing the hard water blues

A major reason that your toilet may flush slowly is a buildup of scale, due to hard water in the high-pressure jet opening (hole) that's built into the trap of the bowl. This jet is what starts the siphoning action and swirling of water in the bowl, drawing waste down and out of the toilet. You can sometimes use a coat hanger and clear the hole, but if the toilet is too scaled or limed up, it's best to replace the toilet (see Chapter 17).

Chapter 8

Clearing the Main Line

In This Chapter

- Unclogging the main line
- Knowing when to call for help

The *main line* is the passageway for the waste that comes from the toilet and from all of the sink and tub traps. The line leads outside the house to the sanitary sewer or septic system. This chapter tells you how to clear a clog in that main line, which is also called the *sewer line*. Makes sense, right?

When the Clog is Beyond the Fixture

Sometimes, the clog is so far from the toilet or sink drain that you can't reach it with a snake (see Chapter 5 for a description of this plumbing tool). If you've fed the snake through the toilet or sink drain to its full length and still haven't reached the clog, your last resort before calling in the pros is to feed the snake through the main clean-out in the sewer line that leads out of your house.

Removing the clean-out plug

The sewer *clean-out* is a fitting with a removable plug. It's usually located at the base of the *main soil pipe* (a large diameter cast iron, copper, or plastic pipe) where it enters the floor of the basement or takes a 90 degree turn to pass through the foundation wall. The clean-out may also be located in the basement floor.

To keep the waste inside the pipe, this fitting has a removable plug that's screwed into the clean-out fitting. Plastic clean-out plugs usually come out easily, but removing a clean-out plug from a cast iron plumbing systems can be a challenge.

In order to remove the clean-out plug, gather up the following tools:

- Large pipe wrench
- Small can of penetrating oil, such as Liquid Wrench or WD-40
- Hammer
- Cold chisel (a thick, short hexagonal, steel bar tool)
- Bucket
- Pair of work gloves
- Goggles or safety glasses

After you assemble your tools, follow these steps:

1. **Locate the clean-out plug.**

   Look for a round plug with a square lug on it.

2. **With your work gloves on, try to open the clean-out plug with a pipe wrench.**

   Place the wrench in the square tab located in the center of the clean-out plug and turn it in a counterclockwise direction.

3. **If using a pipe wrench doesn't work, apply oil to the joint between the soil pipe and plug.**

   Allow the oil to work its way into the joint for 10 or 15 minutes, and then give the pipe wrench another try. This usually loosens a brass plug.

If the plug isn't brass, it probably has rusted into place — in order to get it open, you have to break it into pieces. This is a standard practice with plumbers, but may seem a bit extreme for you to tackle. If you're not comfortable with this, call a professional plumber.

To break up the clean-out plug, follow these steps:

1. **Purchase a new plastic plug so that you can close up the opening after you break up the old plug.**
2. **Put your safety glasses on.**
3. **Use a hammer and cold chisel to break the plug.**

   Place the point of the chisel on the outer edge of the clean-out plug and pound the chisel in a counterclockwise direction (see Figure 8-1). This usually loosens the plug. If not, break off the square tab in the center, and then smash the plug into smaller parts with your hammer.

   Keep in mind that water may be standing in the pipe, just waiting to come gushing out when you crack the plug. Use the bucket to catch any draining water.

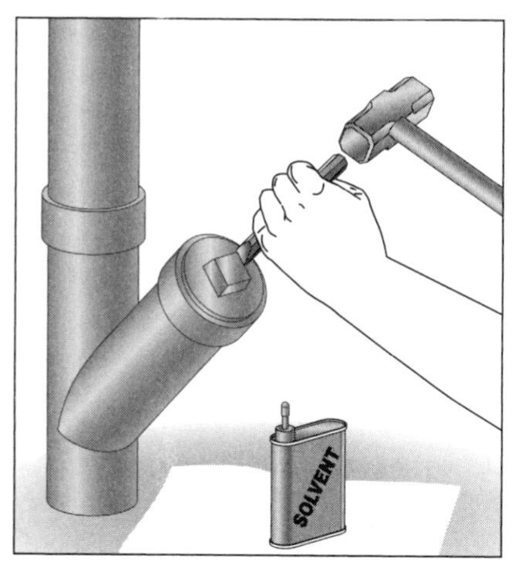

Figure 8-1: Clean out a blockage that's not right at the toilet or fixture by removing the clean-out plug in the main sewer line.

Using a snake to clear the main line

After you get the clean-out open, you can push a snake (see Chapter 5) into the line. Follow these instructions:

1. **Push the snake into the clean out and push it down into the pipe as far as it will go.**
2. **When you reach the clog, keep turning the snake, working it back and forth to loosen the clog (see Figure 8-2).**

   When you feel resistance, you know that you've reached the clog.

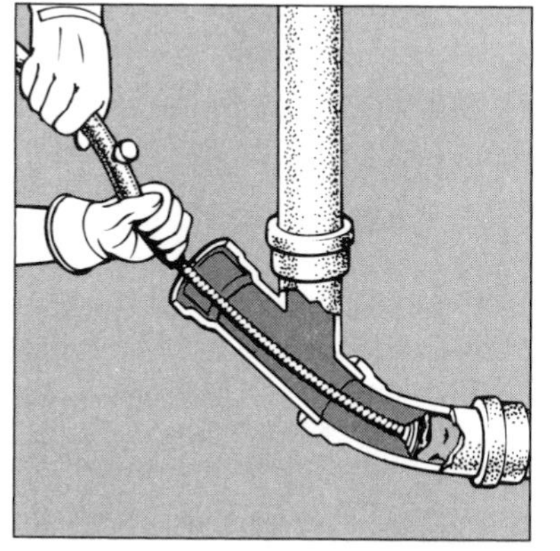

Figure 8-2: Insert the snake into the clean out and feed into drain until you reach the clog.

3. **Run some water through the pipe from a nearby sink.**

   If the water doesn't back up from the clean out, the clog has been cleared.

If the clog isn't clear, see the "Still Clogged? Call a Plumber" section that follows.

4. **Replace the clean-out plug and run hot water into the pipe for several minutes.**

   When you flush the toilet, everything should run okay.

Still Clogged? Call a Plumber

Some blockages are so far down the line that your snake can't reach them. If the blockage is caused by tree roots or some other tough blockage, your little snake won't make a dent. In either case, you could rent a power auger from a rental center, but we don't recommend it. Power augers are difficult to operate and can be dangerous if the end of the steel coil gets lodged in the sewer. This type of machine has changeable cutter heads that are designed to cut roots or auger through tough clogs. If these heads get jammed and the steel coil kinks or breaks, you can get seriously hurt. You've done the hard part by removing the clean-out plug and identifying the problem, but it's time for you to call for help.

Part III
Repairing Leaks Like a Pro

In this part . . .

This part is full of advice about quick fixes for leaking pipes and long-term repairs for all kinds of leaks. We not only give you all of the how-to's, we also fill you in on all of the mysterious jargon that plumbers use — after all, it's pretty cool to be able to describe your problem as a leak in the disc-type cartridge faucet or the shower head diverter valve!

Chapter 9

Fixing Leaky Pipes

In This Chapter

- Making emergency pipe repairs
- Repairing pipe leaks and pipe-joint leaks
- Repairing a frozen pipe
- Quieting noisy pipes

No matter what kind of material your plumbing pipes are made from — threaded galvanized steel, copper (rigid or flexible), plastic, or cast iron — they all can leak. Even a slow, insignificant drip may foreshadow a more pronounced leak down the road.

Pay strict attention to any signs of wetness in your home. A musty odor, moisture, or a stain may indicate that a pipe is leaking behind a wall or ceiling — where you can't see it.

Before attempting any plumbing repairs, turn off the water supply lines so you can work on the pipes without water flowing through them (see Chapter 2).

Don't forget to tell the family that you plan to work on the waste pipes. The last thing you need is someone flushing a toilet and have nasty water running all over you. Place a masking tape "X" across the sink and duct-tape the toilet seat to the bowl as a not-too-subtle reminder.

Temporary Fixes for Pipe Leaks

Sometimes, you can make temporary repairs for pipe leaks — leaks along the length of a pipe are the quickest and easiest to fix. You can buy kits or you can make you own repairs using stuff from you workshop. These fast fixes have been known to last for years — some for over a decade — but don't rely on them to last that long. They're not designed to be permanent repairs.

C-clamp fix

By using a *c-clamp* (which is a steel clamp shaped like a "C" that can be tightened down on the pipe by turning a screw), a piece of rubber, and a small piece of wood, you can temporarily repair small pinhole leaks — even a leak that's squirting all over the place, as long as you don't mind getting wet.

Any piece of rubber will do. (A piece of auto or bike inner tube is perfect, but inner tubes are getting scarce today. You can use *rubber gasket material,* sold at auto parts stores and at most large home centers in their auto parts department.) You also need a couple of pieces of thin wood to cover the rubber patch — try cutting the wood from a paint stirrer.

After you gather up your materials, follow these steps:

1. **Turn off the water (see Chapter 2) and dry off the pipe in the location of the leak.**
2. **Place the rubber patch over the split or pinhole in the pipe.**
3. **Position the wood strips over the patch, and on the opposite side of the pipe.**
4. **Position the c-clamp over the wood strips and tighten the clamp, as shown in Figure 9-1.**

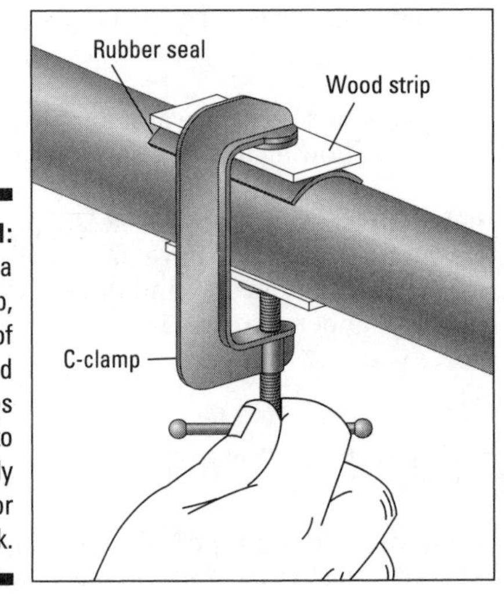

Figure 9-1: Use a c-clamp, a piece of rubber, and some pieces of wood to temporarily stop a minor leak.

If you're repairing a copper or rigid plastic pipe, be careful not to overtighten the clamp — you may crush the pipe. See the following section for more ideas on making temporary repairs on copper or plastic pipe.

5. **Turn on the water to test for leaks.**

   Tighten the clamp a bit more if the patch continues to drip.

Hose clamp fix

Another quick, temporary way to stop a small pipe leak is with a hose clamp. A *hose clamp* is a stainless steel band with notches or vertical holes in it, which make it adjustable. To fasten a hose clamp around a pipe (or any round object), tighten the nut in its closing mechanism with a screwdriver. This type repair is easy to make and works well on copper or plastic pipe, because you can tighten the hose clamp without the danger of crushing the pipe.

Use this repair to try to fix a split caused by a frozen pipe, where the expanding frozen water has split the pipe for several inches instead of creating a pin hole leak. You can apply several hose clamps over a large rubber patch to temporarily stop the waterfall caused by a frozen pipe. See the "Thawing and Repairing Frozen Pipe" section, later in this chapter, for tips on thawing frozen water pipes.

To stop the leak with a hose clamp, gather up at least two hose clamps and a piece of rubber that's large enough to fit over the split in the pipe. A piece of old garden hose, split along its length to fit over the pipe, also works in place of the rubber.

Follow these steps:

1. **Turn off the water and dry off the pipe in the location of the leak.**
2. **Place the rubber patch over the split or pinhole in the pipe.**
3. **Fit the hose clamp over the rubber patch (around the pipe) and tighten it.**

   To loosen the hose clamp in order to fit it over the pipe, turn the screw in the hose clamp counterclockwise to fully open the clamp.

4. **Tighten the clamp by turning the screw in a clockwise direction — see Figure 9-2.**
5. **As the clamp begins to tighten around the pipe, double-check that the rubber and clamp are over the split or pinhole leak.**

   If you need to apply several clamps to a split, position the rest of the clamps before you fully tighten all of them.

6. **Turn on the water to test for leaks.**

   Tighten the clamp a bit more if the patch still drips.

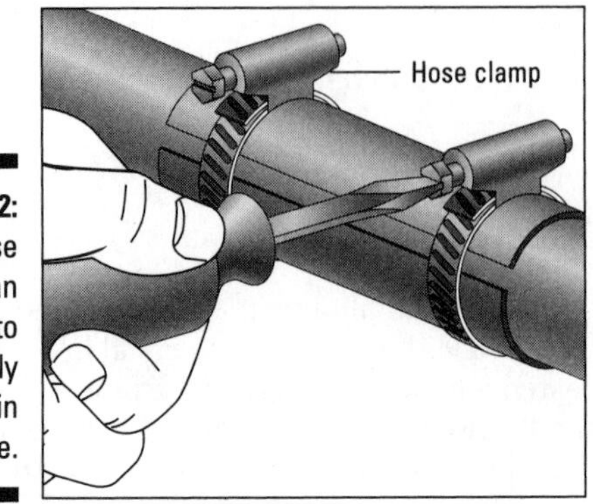

Figure 9-2: Hose clamps can be used to temporarily fix a split in a pipe.

Pipe clamp fix

Instead of a creating homemade patch, you can purchase a pipe clamp repair kit that includes a wide clamp with two semi-circular sides that match the diameter of the pipe you need to mend and a rubber patch. You insert the rubber patch between the pipe and the clamp — when the clamp is tightened, the leak stops. The main advantage to the pipe clamp is that it comes in several different sizes and can be used on a wide range of pipe sizes.

A pipe clamp fix is more permanent than using a c-clamp or hose clamp; however, however you can't considered it a permanent fix.

To stop the leak with a pipe clamp kit, follow these steps:

1. **Turn off the water and dry off the pipe in the location of the leak.**
2. **Place the rubber patch over the split or pinhole in the pipe.**
3. **Open the clamp and place it over the patch.**

   Make sure the rubber patch is positioned over the leak.
4. **Tighten the clamp to compress the patch and seal off the leak, as shown in Figure 9-3.**
5. **Turn on the water to test for leaks.**

   Tighten the clamp a bit more if the patch still drips.

Quick fix for pipe-joint leaks

For a simple leak at a joint, you can use a quick fix, but it will be temporarily at best: applying *two-part epoxy putty,* sold in hardware stores and home centers, around the leaky joint. Pipe joints pose a special problem because the surface isn't smooth and level the way a pipe is. With galvanized pipe, you have the added problem of rusted steel, which keeps the putty from making a tight bond. If the leak is in a drainpipe, however, where there is little water pressure, this repair method is pretty effective.

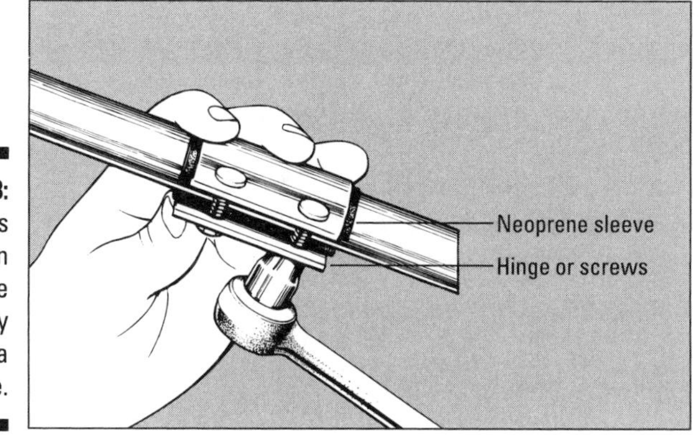

Figure 9-3: Pipe clamps are an effective temporary fix for a leaking pipe.

Before you begin, carefully read the directions on the package of the putty and follow these steps:

1. **Turn the water off.**
2. **Allow the leaky joint to dry thoroughly before you apply epoxy putty.**

   This may present a challenge if the pipe is in a damp location. You can speed up the drying by heating the fitting with a propane torch, heat gun, or hairdryer. Just warming up the fitting is all that's necessary. Allow it to cool before you apply the epoxy patch.
3. **Mix the two-part putty as directed and apply it to the joint, as shown in Figure 9-4.**
4. **Before the putty sets, wrap the pipe in plastic electrical tape.**

   Plastic electrical tape is available at any hardware store or home center.
5. **When the putty is thoroughly dry, turn the water back on.**

The repair may or may not work, depending on the intensity of the leak and the pressure in the line. If the repair doesn't work, call a plumber.

If you repair a joint leak with epoxy putty, check the spot regularly. Epoxy putty is much more likely to fail over time than a clamp repair that's made in a straight section of pipe.

CALL A PRO

Sure signs of future problems

Even though you can fix a leaking pipe, the leak itself can be a sure sign of major plumbing problems down the road. Call your local plumber after you have stopped the drip.

Leaks or leaks in the joints of your galvanized pipes are sure sign that you will have plumbing problems in the future. Galvanized steel pipes have a long life expectancy, but certain water conditions, such as very hard water, can shorten their lifespan. (If you discover several leaks in your galvanized pipes, better sock some money away because before long, you'll probably have to replace the piping through the house.)

Many leaks in threaded-pipe — the old fashioned, galvanized steel-type pipe found in most older homes — are difficult to fix and are often better left to a professional plumber. To permanently fix a whole house full of leaky threaded pipe, you need many lengths of pipe that are cut to precise lengths and threaded on a threading machine, along with a universal joint and a collar (a coupling or elbow that matches the leaking one). Threaded pipe is a complete system: You can't just unscrew a single piece. The pipe has to be cut by a professional and these new parts used for the replacements.

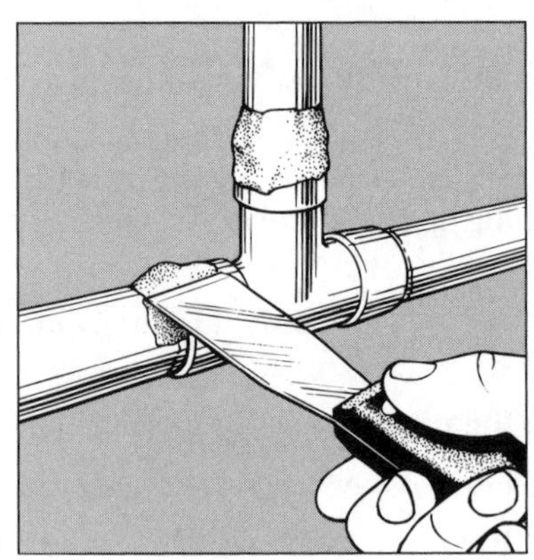

Figure 9-4: Apply the epoxy putty to the pipe joint with a putty knife to temporarily stop a leak in a joint.

Permanently Repairing a Leaky Water Pipe

Leaks in water-supply pipes are usually readily apparent. The water doesn't seep; it sprays. If it's in an open area, you notice a wet wall and a pool of water. If it happens inside a wall, you find wet drywall and water seeping out of the wall along the floor or in the ceiling.

The following sections give you a rundown of the permanent repair techniques used for leaks in pipes of all persuasions — galvanized, copper, and plastic.

Leaky galvanized pipes

Rust is the primary reason that galvanized pipe leaks — and alkaline water promotes rust. The first warning sign is a plugged strainer on faucet aerators (see Chapter 5). If you have to clean the strainers repeatedly and you find bits of rust in the debris, you know that a leak is on the way. This doesn't tell you where the leak is, of course, but to find it, look for telltale rust-colored corrosion around joints.

For a short-term fix, see the emergency repairs in the "Temporary Fixes for Pipe Leaks" section, earlier in this chapter. A pipe clamp or a glob of epoxy putty can stop a leak, but this type of repair is a temporary fix at best. While thousands of pipe clamp fixes have outlived the person who applied the clamp, it still is only a temporary fix.

Replacing a bad section of galvanized pipe can be a little tricky, but if you have a couple of pipe wrenches and a hacksaw handy, give it a try. If you attempt the repair and it's not successful, you can always call a plumber. If this happens, you haven't wasted your time because at least the old pipe will be removed ready for the new one.

Locating a new piece of galvanized pipe

To permanently repair a leaking pipe, you have to remove the bad section of pipe and replace it from elbow to elbow. First, however, you must locate a source for new pipe. Galvanized pipe must be cut to length and then threads must be cut in the end of the pipe. Most home centers and large hardware stores are happy to do this for a small fee.

Before heading out to the store, measure the length of the pipe that you have to replace — measure the distance from the edge of the fitting to the edge of the opposite fitting. Most home water piping has a ½-inch inside diameter so the pipe will extend into the fittings about a ½-inch.

You also need to purchase a union that you'll place in the center of the replacement pipe. The *union* allows you to thread the new piece of pipe into the existing fittings: The pipe has to be turned in a clockwise direction to tighten it, and as you thread pipe into one fitting, it turns in the opposite direction at the other end. The union in the center allows both pipes to be tightened in the fitting and then joined together to complete the project.

The person at the store can cut two pieces of pipe, thread them at both ends, and then assemble the pipe and union for you. When you pick up the new assembly, double-check the overall length of the replacement pipe and union. The whole assembly, after the pipe is fully threaded into the union, should be 1-inch longer than the distance between the fittings that you're removing the old pipe from.

Replacing the piece of galvanized pipe

To replace a section of leaking pipe follow these steps:

1. **Shut off the water to the leaking section.**
2. **Use a hacksaw or pipe cutter to cut the leaking pipe in half, and then unthread each piece from the pipe fittings — see Figure 9-5.**

   Use one pipe wrench to hold the fitting and the other to loosen the pipe. Turn the pipe counterclockwise and apply equal pressure with the other wrench to prevent the fitting form turning.

   Be sure not to loosen the good part of the pipe when you repair the leak.
3. **Open the union and separate the new sections of pipe that you had made at the hardware store.**
4. **Apply pipe dope or Teflon tape to the threads on the end of each pipe section and insert it into the existing fitting.**

   Use two opposing pipe wrenches to prevent the fitting from moving as you tighten the pipe.
5. **Align the sections of new pipe and tighten the union to complete the repair, as shown in Figure 9-6.**
6. **Turn on the water to check for leaks.**

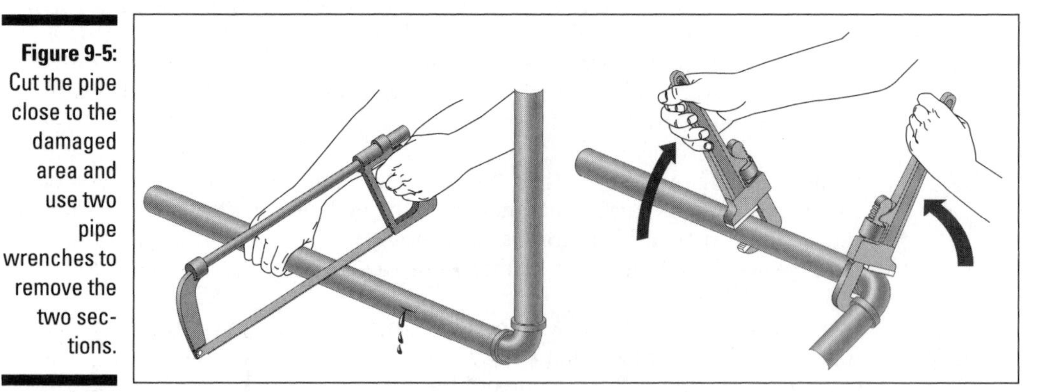

Figure 9-5: Cut the pipe close to the damaged area and use two pipe wrenches to remove the two sections.

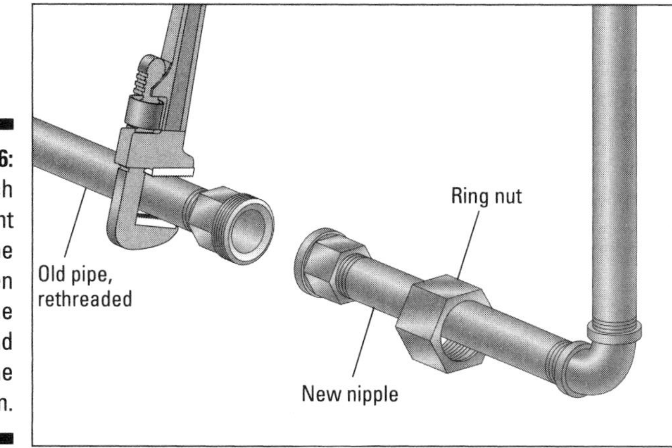

Figure 9-6: Thread each replacement pipe into the fittings, then align the pipes and tighten the union.

Leaky copper pipes and fittings

Typically, people think that with copper pipes in the house, leaks are a thing of the past. Not true. Sulfur and excessive oxygen in the water can cause copper pipes to fail prematurely. Pinholes appear, with a resulting spray of water.

Repairing leaks in copper pipes

Repairing a copper pipe is easier than repairing galvanized pipe. You simply cut out the bad section of pipe with a hacksaw or tubing cutter (see Chapter 3), and solder in a new section of pipe (covered in Chapter 4). The secret to this repair is using a pair of *slip couplings* — fittings that look like standard couplings that are used to join two pieces of pipe, but without the flange in the center, so a slip coupling slides completely over the end of the pipe.

Most leaks in a copper pipe are probably only a pinhole, but you still want to replace at least six inches or more of the pipe. You can purchase a piece of copper pipe and two slip couplings at most home centers or large hardware stores. You also need a propane torch and solder (see Chapter 3).

To repair a pinhole in a copper pipe, follow these instructions.

1. **Turn the water off and if possible, open the hot and cold faucets in the basement.**

   Also open some faucets on the upper floors to allow air into the system, so the water drains out of the pipes.

2. **Use a hacksaw or tubing cutter to remove the leaking section of pipe.**

   Have a bucket ready to collect any water remaining in the pipe that will run out of the first cut.

Stopping the drip of a compression valve

Fittings and valves connected to flexible copper pipe are subject to bumps, so they can leak. Tightening the compression nut slightly (about ¼ turn) can usually stop the leak — see the figure below. The leak may also come from the valve — tightening the packing nut around the valve stem should stop this leak. If none of these quick-fixes work, replace the fitting or valve with a new one (see Chapter 4).

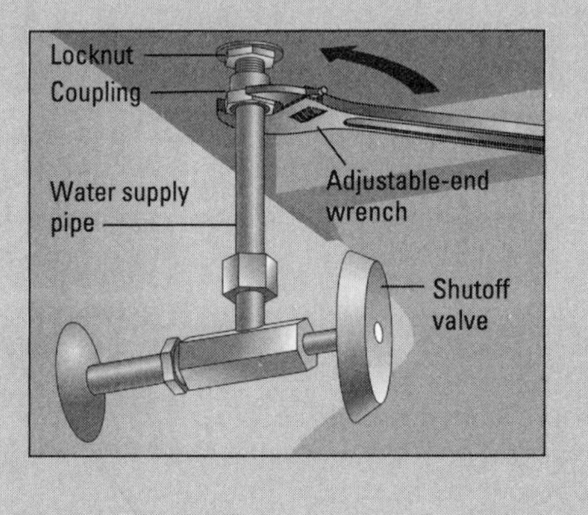

3. **Make a second cut on the other side of the leak and remove the bad section.**
4. **Clean each end of the pipe, and the ends of the short replacement pipe.**

   Also clean the inside of the slip couplings. Apply flux to all pipe ends and to the inside of the slip couplings. (See Chapter 3 for more about flux.)
5. **Slide the slip couplings completely onto the new piece of pipe, and place the pipe in the gap left by the pipe that you cut away. Slide the slip couplings onto the old pipe. See Figure 9-7.**
6. **Solder the new piece of pipe in place (see Chapter 4).**
7. **Close the faucets. Turn the water on to test for a leak.**

   Open the upstairs faucets to let the air out of the system, and you're back in business.

Watch out for any water that may drip out of the pipe while you're soldering — it's boiling hot. Also, solder won't adhere to pipes with water on them.

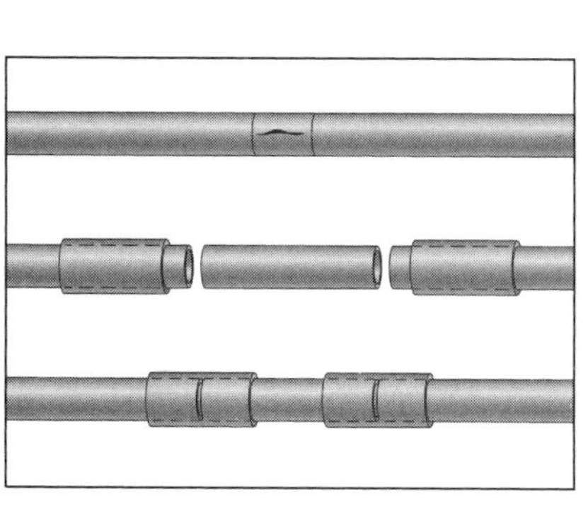

Figure 9-7: Push the slip couplings on to both ends of the new pipe and put the pipe in place. Then slide couplings over the old pipe and solder.

If getting all of the water out of the line is a problem, and small amounts of water keep dripping out of the pipe, soldering the replacement piece will be difficult. In this case, use a copper union. Solder the union nuts to the end of one section of pipe. (Don't forget to slip the union nut over the pipe before you begin soldering!) Solder the other half of the union to the short piece of copper pipe. Then, solder the pipe to the existing section with a slip coupling. The open union allows the water to drip out while you're soldering the repair together.

Repairing leaks in copper fittings

Leaks can also occur at a poorly soldered fitting. You may think that if the joint doesn't leak initially, it'll be okay for the long run, but sometimes a very thin line of solder in the joint breaks because of a combination of pipe movement and water pressure. Repair is a matter of resoldering the joint. You need to remove the fitting, clean the joints, and solder it again. (See Chapter 4 for soldering and unsoldering tips.)

If you have particularly hard water, check with your local building department. They can give you information as to the condition of the water, because chemicals in the water may cause a pinhole leak. In this case, installing a union or coupling at the hole may temporarily fix the problem, but the rest of the pipes will probably eventually fail. Using a water conditioner may help prevent further damage to the pipes. In situations where the water is particularly aggressive (that is, very hard), you may want to replace the copper with plastic (CPVC) or PEX lines (see Chapter 4) rather than using a water conditioner. This pipe replacement is definitely a job for a plumbing contractor.

Leaky plastic pipes

Water shouldn't cause any deterioration of plastic pipes. Most pinhole leaks are caused by mechanical damage, such as driving a screw or nail into the pipe. You'll know it immediately. Repair is easy (unless the pipe is hidden behind a wall). You cut out the bad section and use the same type slip couplings to reinstall the new section of pipe. (See Chapter 4 for tips on working with plastic pipe.)

To repair a pinhole leak in a plastic pipe, follow these instructions.

1. **Turn the water off and if possible, open the hot and cold faucets in the basement.**

   Also open some faucets on the upper floors to allow air into the system, so the water drains out of the pipes.

2. **Use a hacksaw or tubing cutter to remove the leaking section of pipe.**

   Have a bucket ready to collect any water remaining in the pipe that will run out of the first cut.

3. **Make a second cut on the other side of the leak and remove the bad section.**

4. **Clean each end of the pipe, and the ends of the short replacement pipe.**

   Also clean the inside of the slip couplings.

5. **Apply CPVC cement (see Chapter 3) to all pipe ends and to the inside of the slip joint fittings.**

6. **Slide the slip couplings completely onto the new piece of pipe, and place the pipe in the gap left by the pipe that you cut away. Slide the slip couplings on to the old pipe.**

7. **Close the faucets. Turn the water on to test for a leak.**

   Open the upstairs faucets to let the air out of the system, and you're back in business.

We occasionally experience a few failures where the glued joint has failed, but using pipes and fittings from different manufacturers, which results in a poorly welded joint, usually causes this. These failures typically occur shortly after the piping has been installed — in other words, you'll know it right away. If it leaks, replace the fitting with one that fits.

Fixing a Leaky Drain Pipe

Leaks in cast iron pipe are almost always of the seeping variety around the joints. Three- and four-inch pipe is used to drain toilets and provide a connection to the main sewer or septic system. This system seldom leaks, but if trouble does occur, sections of these pipes can be cut out and replaced.

Leaks in drain pipes are relatively rare and except for fixing simple under-sink leaks and minor repairs to PVC drainage systems — in traps under the sinks and at shower drain joints — making repairs to cast iron or galvanized steel pipe is best left to the professional. To make minor repairs on sink traps and shower drains, refer to the two following sections. In Chapter 12, you can find more information on repairing toilet seal leaks.

Cast iron pipe is heavy and must be properly supported. This is why we don't recommend doing this project yourself. If you remove a section of the pipe without proper support, the rest of the stack may come crashing down. Contact a plumber for a job of this type.

Fixing sink trap leaks

The drain pipes under sinks are subject to a lot of bumps and bangs, and the slip joints may begin to let leak a little. The parts of the trap assembly under your sink have particular names — see Figure 9-8.

- The pipe attached to the underside of the sink is called the *tailpiece.*
- The U-shaped pipe connected to the tailpiece is called the *trap* and is connected to an *elbow* that protrudes from the wall.
- At each joint, there is a *rubber washer* that's compressed around the pipe to prevent leaks.

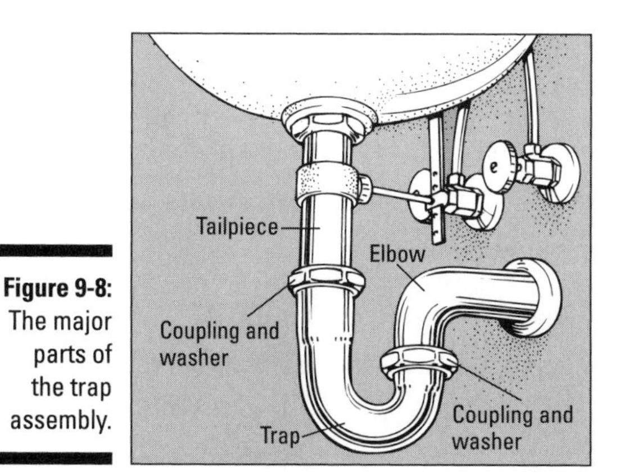

Figure 9-8: The major parts of the trap assembly.

The rubber parts harden over time and the joints can suddenly start to leak, but usually, they begin to leak after they've been disturbed for some reason. If you have an old sink and the trap isn't leaking, be careful that you don't hit the trap assembly while storing something under the sink.

The same is true for older metal parts of the trap. Don't be too hard on an old sink trap when trying to unclog it. The pipes may be rusted in spots and ready to leak at the slightest nudging.

If the parts of the trap assembly are chrome plated and appear to be in good shape, simply replace the rubber washers. If the parts don't look good, take the parts to the store and purchase a whole new replacement trap assembly. If you currently have metal drain pipes that aren't exposed to extreme temperatures and are not visible under the sink, replace them with plastic — plastic doesn't rust.

Loosen the slipnuts that hold the tail trap to the tailpiece, and then remove slip washer that holds the trap to the elbow. Pull the trap off the tailpiece. Remove the elbow from the drainpipe by loosening the slipnut holding it in place.

Thawing and Repairing Frozen Pipe

Before you can thaw a frozen pipe, you need to locate the frozen area. Check the water flow from several faucets throughout the house to help you isolate the problem. To find the frozen area, inspect the pipe, looking for cracks or splits and expanded or deformed areas. Different types of pipes crack in different ways:

- Galvanized pipe tends to split.
- Copper either splits or changes diameter.
- Plastic pipe cracks in long sections.
- Polybutylene (PB) usually expands. Polybutylene occasionally breaks, but freezing problems are minimal compared with other types of pipe. Polybutylene is also flexible enough that you can gently squeeze the pipe with pliers to find the frozen areas. If it's frozen, the pipe is hard and inflexible.

After you find the frozen pipe, follow these steps to thaw it.

1. **Turn off the main water-shutoff valve (see Chapter 2).**

   If you don't do this, you may have water spraying through the broken pipe as the ice melts.

2. **Open the fixture or faucet closest to the frozen pipe.**

   This allows any steam you may generate by thawing the water in the pipe to escape.

3. **Using a heat gun, apply the heat over a large area, moving it slowly over the entire length of the pipe, as shown in Figure 9-9.**

   The idea is to *gradually* warm the entire pipe, so don't hold the heat source at one location. Be patient; the pipe will thaw in time. And doing so slowly reduces the danger of damage to you or to the plumbing. You can use a blow dryer, but it takes considerably longer to unfreeze the water inside the pipes.

Figure 9-9: Use a heat gun to thaw a frozen pipe. Keep the gun moving to apply heat to a wide area to prevent pressure build up in the pipe.

Sometimes a quicker (and safer) way to repair the frozen pipe is to cut it out and replace it with a new pipe. (See the "Permanently Repairing a Leaky Water Pipe" section, earlier in this chapter, for information on repairing pipe.)

Quieting a Noisy Pipe

If you have a noise that sounds like a jackhammer in the walls every time you turn a faucet on and off, you have what's commonly called *water hammer.* This is what happens: When a faucet or valve opens, water rushes through the pipes to the outlet. Then, when the valve closes quickly, the rushing water comes to a sudden stop. The energy of the rushing water is transferred into the metal pipes and it sounds as if someone is hitting the pipes with a hammer. Clothes washers and dishwashers are the main culprits, because their electronically controlled valves close very fast. Over a period of time, the constant hammering and vibration can damage water lines and create leaks.

Preventing frozen pipes

If you've ever had to deal with frozen pipes, you know how important it is to try to prevent them. The pipes most likely to freeze are the ones that run through poorly insulated outside walls or between an unheated basement-level garage. So to avoid the problem, follow these preventive measures:

- Keep kitchen or bathroom cabinets on outside walls of a house open during extremely cold weather. This allows the warm air from the room to circulate inside the cabinet and around the pipes.
- Let water dribble in plumbing fixtures whose pipes run through cold spaces because running water does not freeze as readily as still water.
- Use heating cables or electric heating tape designed to wrapped around exposed pipes. The tape contains a thermostat that turns the tape on when the temperature gets close to freezing.

  Heat tapes are a good solution for freezing pipes but should not be used under insulation. Most heat tape manufacturers warn against wrapping the heat tape around the pipe and then placing insulation over it. Also read the installation instructions on the tape some types warn about placing the tape around plastic pipe.
- Wrap the pipes with aluminum-backed insulation or preformed foam insulation to protect pipes for short cold spells.
- Relocate water pipes that are in an outside wall to a heated section of the house.

The most common solution is to install an *air column,* a vertical pipe with a cushion of air in it, near the valves, in order to absorb the pressure of the rebounding water. Most galvanized piping systems have this type of anti-hammer system, shown in Figure 9-10. This works, but after a while, the air is absorbed into the water, the pipe fills with water and the noise will return. The air can be replenished by turning off the water supply, opening all faucets and draining the lines. All of the water must be removed, so that the air chamber can drain and refill with air. It takes time, so be patient.

If your piping doesn't have a series of air chambers, you can install them yourself — several types can be purchased from your local home center or hardware store. You can also make up your own air chambers with short pieces of pipe capped off at the top.

Another option is to use an air-hammer muffler that has a built-in rubber bladder that prevents it from becoming waterlogged.

One of the easiest devices to install is a water-hammer arrestor. These screw-on devices are designed to go between the faucet and the hoses supplying

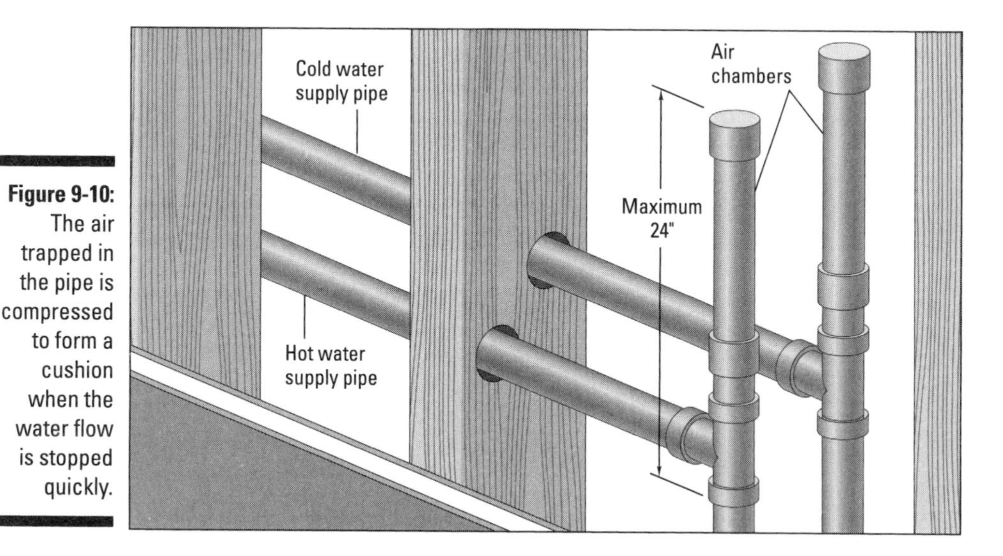

Figure 9-10: The air trapped in the pipe is compressed to form a cushion when the water flow is stopped quickly.

the clothes washer. You can find them in most plumbing sections of home centers or hardware stores. Follow the instructions that come with the device.

The following steps can help you install an anti-hammer device to a clothes washer:

1. **Turn off the water to the hot and cold faucets that supply the clothes washer.**
2. **Detach the hoses that run from the washer to the faucet.**
3. **Thread the anti-hammer chambers onto the faucet.**
4. **Reinstall the hoses on the threaded nipple that protrudes from the device.**

You can also find an anti-hammer device that's designed for under a sink. Although you want to follow the instructions that come with the device, here's a brief rundown of how you install one of these:

1. **Turn off the water to the sink.**
2. **Remove the riser tube leading from the stop valve to the faucet.**
3. **Remove the valve from the pipe protruding from the wall and apply pipe dope or Teflon tape to the end of the pipe.**
4. **Thread the anti-hammer device onto the faucet and reinstall the stop valve on the device.**

   Some of these require that a pipe tee be inserted between the valve and wall supply pipe.

Chapter 10

Fixing a Leaky Sink Faucet

In This Chapter

- Stopping a compression faucet or washerless faucet leak
- Halting a stem leak
- Repairing a ball-type, cartridge-type, or ceramic disk-type faucet
- Fixing a leaky sink sprayer

Stopping a leaking faucet is basic and easy to do. The challenge, however, is first determining what type of faucet you have. After you know that and have the replacement or material to stop the leak, there's not much to it.

Not only do you save water by stopping the leak, but you also get a sense of satisfaction — some may even call it smugness. You did it yourself and the next time you notice a leak, you'll be ready and able to fix it.

Before you begin to repair a leak for any type of faucet, turn off the hot and cold water shut-off valves under the sink (see Chapter 2).

Stopping a Compression Faucet Leak

A *compression faucet* is called this because a rubber washer, actually made from neoprene (a form of synthetic rubber) is forced against a metal seat to choke off the water flow. As you turn the handle, the washer is pushed against (compresses) the valve seat (see Chapter 3). If the washer or valve seat becomes damaged, a seal isn't made and the faucet leaks.

If your compression faucet is dripping, most likely the rubber washer has worn away. Less common is a worn metal seal that the washer presses against when it's closed. The metal (usually brass) valve seat can become damaged if you don't change the washer before it's too worn — metal then grinds against metal and chews up the seat. Hard foreign matter can become trapped between the valve seat and the washer. If this happens, closing and opening the faucet grinds the particles inside, damaging it beyond a simple washer replacement.

Replacing a worn washer

To replace a worn washer, follow these steps, using Figure 10-1 as a guide:

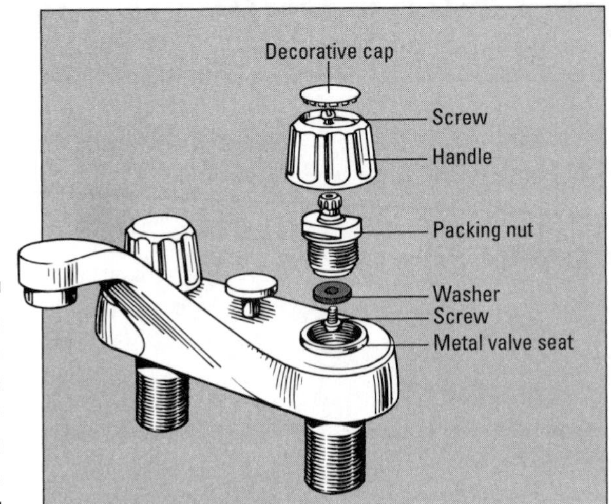

Figure 10-1: Here are the major parts of a compression faucet.

1. **Turn off the water to the faucet.**
2. **Remove the decorative cap, if there is one, on top of the faucet handle.**

   Depending on its design, either pull it up or unscrew it.
3. **Unscrew or pull off the handle and remove it.**

   If the handle sticks, gently nudge it up with a screwdriver. Wrap the screwdriver edge with a rag to prevent marring the finish.
4. **Remove the cover over the valve, called an *escutcheon,* if there is one.**

   Some types unscrew while others are held in place by set screws. Inspect the escutcheon to figure out how to remove it.
5. **Unscrew the packing nut that holds the body of the valve in place, turning it counterclockwise.**

   The valve stem should come out of the base of the valve. You may have to twist the valve-stem body several turns after you loosen the packing nut and the valve stem. If it is hard to unscrew, put the handle back on the stem and give it a twist. The stem then comes out of the valve.
6. **Unscrew the retaining screw and remove the washer.**

On the other end of the stem there is a rubber washer that is held in place by a screw. These valve washers come in many shapes and sizes, so take the valve stem to the plumbing department of your local hardware store or home center, and get a washer that matches the old one. This may be more difficult than it sounds because the old washer is usually damaged and deformed. The best clue to the original shape of the washer is to look into the valve body and take a look at the metal opening that the washer presses against — the valve seat. If the side of the valve seat is angled, replace the washer with a cone shape; if the valve seat is flat, get a replacement washer that's flat.

7. **Replace the old washer with the new one and reassemble the faucet.**

Replacing the valve seat

If the faucet still leaks after reassembling, the seat may be damaged.

To replace a damaged valve seat, follow these steps:

1. **Disassemble the unit — see the "Replacing a worn washer" section earlier in this chapter.**

   While the stem is out, look into the valve body and inspect the faucet seat. (Look for a brass insert inside the valve body that the rubber washer presses against to stop the flow of water.) If this seat is rough, it will tear up the new washer and the valve will begin to leak again. A rough valve seat should be replaced. If the faucet doesn't have replaceable valve seats, the seat can be ground smooth with a valve seat grinder.

2. **If the valve has a removable seat, remove it.**

   If the seat has a hexagonal or grooved opening in its center, remove the seat with a screwdriver or Allen wrench.

   If the seat isn't removable (it will have a round hole) you have a really old faucet. You can grind it smooth with a seat-grinding tool, found in the plumbing department of hardware stores and home centers and shown in Figure 10-2. The tool comes with instructions and is easy to use. It's also known as a *faucet seat reamer.* The tool fits over the valve seat (where the washer usually rests) and grinds the seat with the tool. The idea is to reshape the damaged seat to accept the new washer. Be careful when using this tool to keep it perfectly aligned or the seat will be dressed unevenly.

3. **Take your old faucet to a well-stocked plumbing department or supplier, along with the faucet stem, to ensure you get the correct replacement.**

Faucet seats come in many sizes and with many different thread patterns — a perfect match is critical.

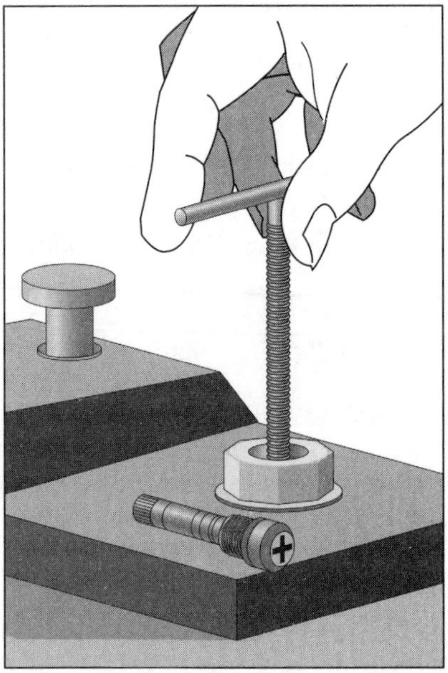

Figure 10-2: Align the valve seat grinder in the center of the valve body then press down on the handle and turn three or four times to grind it smooth.

4. **Replace the valve seat, being careful not to cross-thread the connection.**
5. **Replace the stem washer and packing as necessary.**

   Before you replace the stem, coat the washer and all moving parts — including the handle stem — with heatproof grease. This type of grease doesn't break down in hot water and keeps the stem and faucet working smoothly for a long time.
6. **Reassemble the faucet in the reverse order that you removed it.**

Stopping a stem leak

If you have a leaky handle rather than a drippy faucet, the water is leaking past the *stem packing* — see Figure 10-3 — or the washer. Older faucets have a string-like substance wrapped around the handle stem to hold the water back. The packing eventually wears and water can sneak between the stem and packing. Newer faucets stop the water leak with an o-ring or washer.

Older faucets with packing are most likely to leak. Here's how to fix them:

1. **Turn off the water.**
2. **Remove the handle from the shaft.**

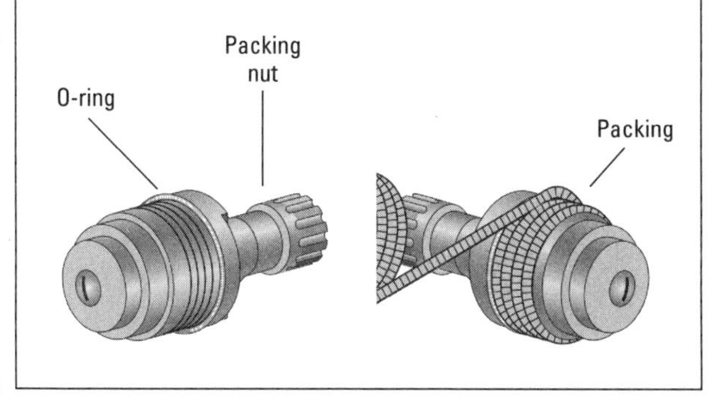

Figure 10-3: Replace the packing around the valve stem and re-tighten the packing washer.

3. **Tightening the packing nut.**

   Turn it clockwise about a ½ turn. This may be all that is necessary to stop the leak.

4. **If tightening the packing nut doesn't work, loosen the packing nut with slip-joint pliers or a wrench, and then unscrew the nut by hand and remove it.**
5. **Remove the old packing from around the stem.**
6. **Replace the old with new packing.**

   You can find new packing in the plumbing department of hardware stores and home centers.

7. **Reassemble the faucet.**

Keep a supply of various sized o-rings, packing washers, packing rope, and washers handy to save you a trip to the store at the first sign of a leak.

Stopping a Washerless Faucet Leak

The new variety of washerless faucets is easier to fix than compression faucets. The hardest part of this project is figuring out what type of washerless faucet you have. The three following sections explain how to fix the three most-common types of washerless faucets currently on the market.

The best tip-off as to what type faucet you have is how the handle that controls the water moves.

- If the control handle moves in an arch (up and down and sideways), is attached to a dome-like top of the faucet, and has a small set screw in the base of the control handle, then you have a *ball-type faucet.*
- If the handle seems to move up and down to control the water flow and then right or left to control temperature, you have either a *cartridge-type faucet* or a *ceramic disk-type faucet.*

If you can find the user's manual that came with the faucet (small miracle) you may find the manufacturer's name and model number. Most manufacturers sell a repair kit and these are usually stocked by home centers and large hardware stores. These kits have all the necessary parts, including any special tools needed to take the faucet apart and repair it.

Ball-type faucet

Follow these steps to fix a single-handle ball-type faucet, shown in Figure 10-4.

1. **Turn off the water.**
2. **Remove the handle.**

   Loosen the set screw that secures the handle to the shaft coming out of the ball valve. The screw head is on the underside of the lever. This set screw requires an Allen wrench to loosen. (An Allen wrench is an L-shaped hex wrench that fits into the recessed socket in the head of the set screw. You can purchase a set of Allen wrenches at any hardware store or home center in the hand tool department.) Some faucets are manufactured in Europe and may have metric sized parts. If you can't seem to find an Allen wrench that fits the set screw, chances are the screw has a metric head. You need to purchase a set of metric Allen wrenches — they aren't expensive.
3. **Remove the ball valve and spout.**

   Wrap tape around the jaws of your wrench or slip-joint type pliers to protect the valve parts. Loosen the *cap assembly* (dome-shaped ring at the top of the faucet) by turning the adjusting screw counterclockwise. Grab the shaft that the handle was attached to, move it back and forth to loosen the ball valve assembly, and then pull it straight up and out of the faucet body.
4. **Replace valve seats and o-rings.**

   When you look inside the faucet body, you see the valve seats and rubber o-rings, and behind them are springs. Remove the seats, springs, and o-rings from the faucet body and take them to the hardware store or home center to be sure you get the correct repair kit.

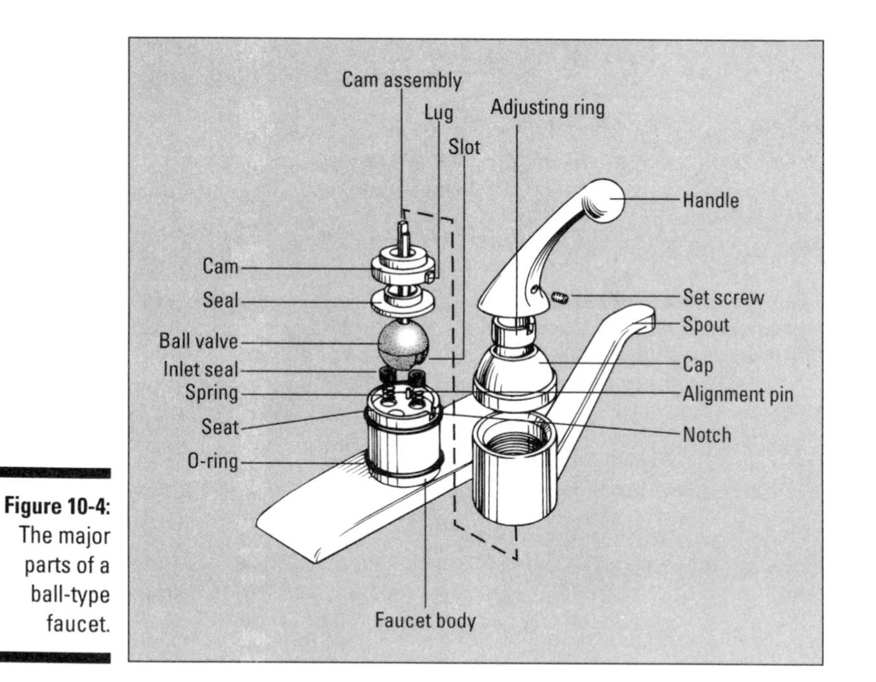

Figure 10-4: The major parts of a ball-type faucet.

5. **Replace the parts and reassemble the faucet in the reverse order that you took it apart (or follow the directions in the repair kit).**

   Pay attention that you reinstall the ball in the same position as you removed it.

6. **If a faucet leaks around the handle or the spout when the water is running, tighten the adjusting ring.**

   Under the handle is an *adjusting ring* that's screwed into the valve body. Slots in the top edge of the ring allow you to insert the adjusting tool into these ring and turn it.

   If you can't find the adjusting tool, use a large screwdriver or slip-joint pliers to turn this ring in a clockwise direction to tighten it. Unless the ring is very loose, tighten it only about ⅛ turn.

7. **Turn the water on and slip the handle back on the control ball's shaft. Adjust the ring so the leak around the ball shaft stops but the ball can be easily adjusted.**

   If you can't get the leak to stop, the seal under the adjusting ring is bad and should be replaced.

8. **Tighten the set screw to secure the handle.**

Cartridge-type faucet

Follow these steps to fix a single-handle cartridge type faucet, shown in Figure 10-5:

1. **Turn off the water.**
2. **Remove the handle.**

   Remove the cover on the top of the handle to expose the screw that holds the handle to the valve stem. Pop off the cover by placing the tip of a screwdriver between the cover and the handle housing and prying up. To remove the handle, turn the screw in the center of the cap counter-clockwise; remove it then pull the handle up and off the valve assembly. This exposes the valve stem coming out of the valve cartridge.
3. **If your faucet has a movable spout, remove the pivot nut.**

   Use an adjustable wrench or slip-joint pliers to loosen (turning counter-clockwise) to remove the pivot nut at the top of the faucet body. This nut holds the spout sleeve in place and prevents water from coming out the top of the faucet.
4. **Remove the spout assembly by twisting it back and forth as you pull up.**

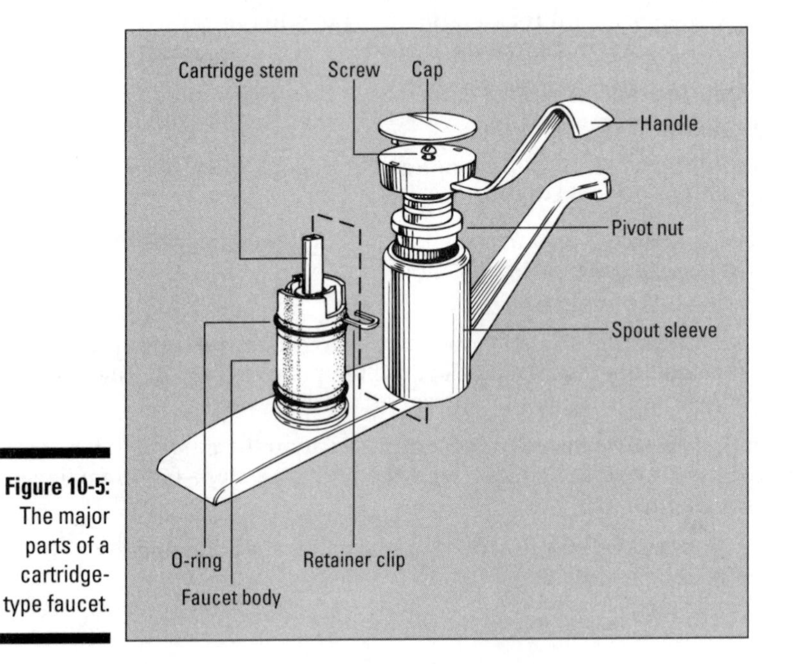

Figure 10-5: The major parts of a cartridge-type faucet.

5. **Remove cartridge clip and replace cartridge.**

   To remove the cartridge, you have to pull out the small U-shaped clip that holds the valve cartridge in the faucet body. Pry the clip loose by placing the tip of your screwdriver between the faucet body and the U-section of the clip.

   Twist the screwdriver and the clip comes out. Grab the clip with your pliers and remove it. Pull up on the cartridge stem with a twisting motion. If it doesn't twist out easily, reinstall the handle so that you can get a good grip on the shaft to pull the cartridge out. Take the old cartridge to your hardware store or home center and purchase a replacement kit.

6. **Reassemble the faucet according to the directions.**

Be sure to replace the cartridge in the correct position — there may be a notch in the valve body that the cartridge fits into. How the cartridge is inserted into the valve body determines which side the hot and cold water are on. Usually you move the lever right for the cold and left for hot. If you reverse the position of the cartridge, the hot and cold will be on opposite sides. If this happens, take the faucet apart and reverse the cartridge position.

Ceramic disk-type faucet

A ceramic disk-type faucet is reliable and usually doesn't require much maintenance. They can be a bit tricky to take apart, however. Older models are held together by screws underneath the faucet, so if you can't figure out how to get the handle off, look under the counter and you should see a couple of brass screws. Loosen these and the whole cover and handle will come off the faucet, revealing the valve cartridge — see Figure 10-6.

1. **Remove the handle.**

   Lift the handle to its highest position to expose the set screw holding it in place. Use an Allen wrench or screwdriver to turn the set screw in a counterclockwise direction, then lift off the handle. With the handle off, pull the decorative trim cap up and off the cartridge body.

2. **Remove the valve seals and cartridge assembly.**

   Loosen the two screws on the top of the valve cartridge and lift the assembly off the faucet body.

3. **Replace the rubber seals.**

   You can find several rubber seals under the cartridge — replacing them stops most leaking. If you replace these seals and the faucet still leaks, the valve cartridge, which contains the ceramic disks, is worn and must be replaced.

4. **Take the cartridge assembly to your local hardware store or home center and purchase a replacement kit.**
5. **Install the new cartridge according to the instructions in the kit.**

   The kit will contain new o-rings to seal around the faucet body. Be sure to replace these even if the old ones look like they're in good shape.

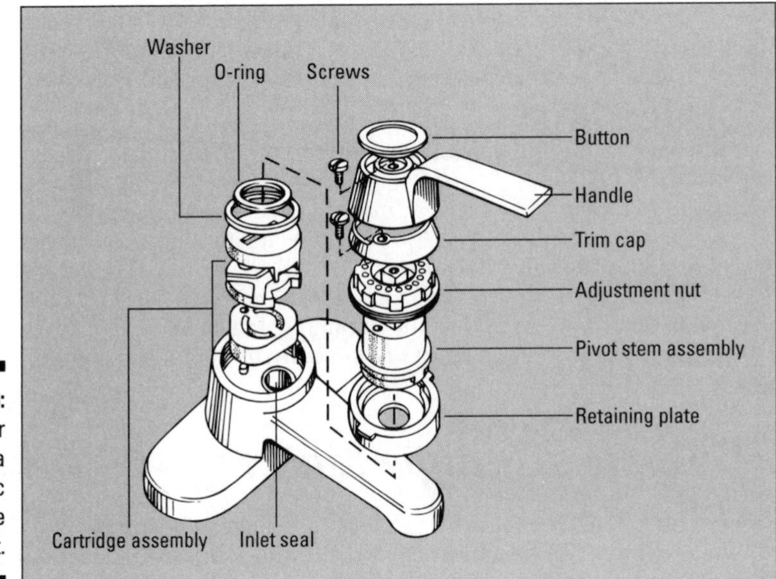

Figure 10-6: The major parts of a ceramic disk-type faucet.

Fixing a Dish Sprayer

If you notice that the water flow out of the dish sprayer isn't what it used to be, you can't use this as an excuse not to do the dishes. The fix is easy; just as the aerator may become stopped up once in a while (see Chapter 5), so can your dish sprayer.

Here's how to clean the spray head (see Figure 10-7):

1. **Turn the water off from under the sink then remove the spray head.**

   Untwisting the head disassembles some models; other types are held together by a screw.
2. **Clean out any blockage in the small holes of the spray head.**
3. **Check the spray hose for kinks.**

Look under the sink and check the condition of the spray hose. It may become entangled with objects stored under the sink and kink, restricting the water flow. Replace a badly kinked hose. You can purchase replacement sprayer assemblies at most home centers, but take the hose and spray head with you when you shop for a replacement: Turn the water off and then unscrew the hose from the base of the faucet.

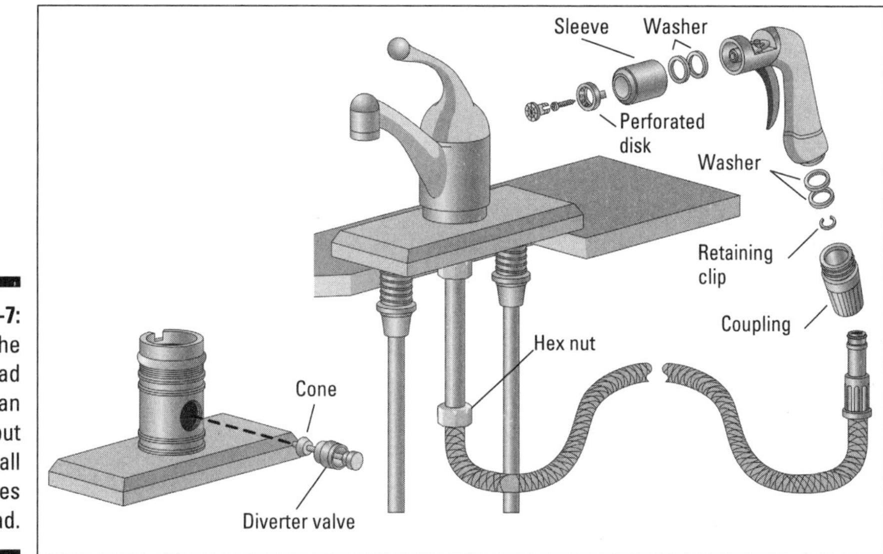

Figure 10-7: Remove the spray head and clean deposits out of the small spray holes in the head.

4. **Check the diverter valve.**

   If the spray head and hose check out okay, a malfunctioning *diverter valve* may be the problem. It's a very simple device that's activated by water pressure. When the sprayer is off, the water is diverted to the spout. When you press the sprayer trigger, the water pressure in the hose drops, and the valve closes off the water flow to the spout and directs water to the sprayer. To do this, it must move freely. A sure sign of a bad diverter valve is when water flows from both the faucet and the dish sprayer at the same time. This valve is located in the base of the faucet behind the swivel spout, if there is one. Refer to Figure 10-7 for the location of this part. To service it, make sure that the valve moves freely and that all the rubber parts are in good shape. Replacement parts are available.

5. **Reassemble the sprayer.**

Chapter 11

Fixing a Leaky Tub or Shower Faucet

In This Chapter

- Discovering tips for repairing a two-handled faucet
- Getting to know the types of single-handled faucets
- Fixing a diverter leak
- Repairing your showerhead

Bathtub faucets have just as many reasons to leak as the faucet in your kitchen sink does. The operating parts inside the faucets work the same way, but there is a difference: Bathtub faucets (and the diverter valve that sends water to the showerhead) are hidden inside a wall, which makes them more difficult to repair.

Repairing these leaks can cause the family to get a bit more uptight than with other repairs because you usually have to turn off the water supply to the entire house — most tubs and showers don't have a separate shutoff valve. Turn off the water and send your gang to the movies while you do the repairs.

Repairing a Two-Handled Tub Faucet

If you have a separate handle to control the hot and cold water running into your tub you have a *two-handled faucet.* Older versions of this type of valve use rubber washers to control the water flow; newer versions contains a valve cartridge. The basic repairs to these faucets are covered in Chapter 10.

Every few years, manufacturers of plumbing fixtures come out with a better tub faucet. This means that you can find hundreds of different variations of the same basic model. But like two-handled sink faucets, tub and shower valves all work pretty much the same.

Before you begin the repair, analyze the leak.

- **Does the faucet leak through the spout when it's closed?** If so, the seat washer or valve cartridge need to be replaced.
- **Does the water dribble out of the faucet when the shower is on?** The diverter valve in the spout or in the valve inside the wall is leaking and should be cleaned or replaced.
- **Does water appear around the faucet handles when you turn the faucet on, and does it stop when you turn the faucet off?** The packing seal or o-ring around the valve stem is leaking.
- **Does water leak out of the shower head or squirt all over the place during a shower?** The shower head needs cleaning.

Accessing the valve

The biggest challenge in fixing a leak in one of these valves is getting at it. Unlike a sink faucet valve that's sitting out in the open, these valves are enclosed inside the wall. To repair one of these valves you have to figure out how to remove the handles and the *escutcheon trim* (cover over the valve) that surrounds the handle stem.

If this is the first time that you've had this valve apart, you need to find a hardware store or home center that has a good selection of plumbing parts. The washer on the end of the stem must fit perfectly. If you decide to replace the packing, it must be an exact replacement. Older faucets used packing string made of graphite or Teflon, while newer faucets may use nylon or fiber bonnet seals or rubber o-rings.

Before you begin to chip away at the wall to get at that leaky valve, make sure that some nice plumber has not already provided an access panel. Many older homes have a panel in the opposite side of the wall that contains the plumbing. These are sometimes located in a closet, so look for it before you do anything else.

Fixing the drip

Take a close look at the Figure 11-1 to identify the major parts of this type shower valve assembly. Then take a look at your shower valve and follow these general steps to fix that drip.

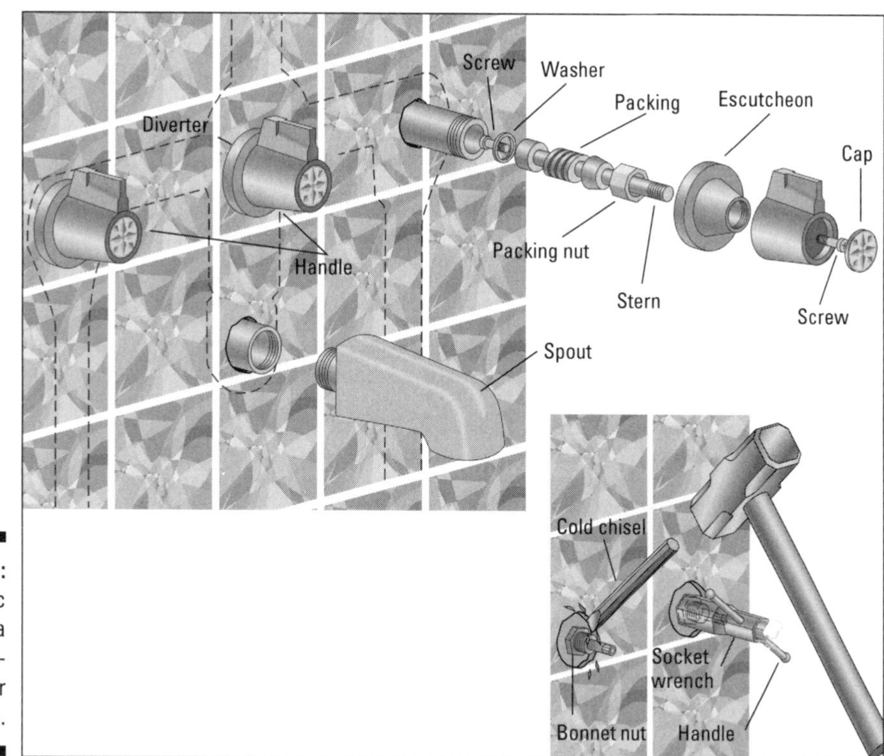

Figure 11-1: The basic part of a compression shower valve.

1. **Turn the water off.**
2. **Remove the handle.**

   Remove the cap on the end of the handle, and then remove the screw that holds the handle to the stem. This screw may be hard to remove if it hasn't been out for several years. Use a screwdriver that fits snugly in the slot or head and hold it tightly against the screw. Now is *not* the time to strip the drive head.

   If the screw resists, clamp locking pliers or an adjustable wrench on the screwdriver shank. While pushing the screwdriver against the screw, try turning the screwdriver. Another method that works well, particularly with Phillips screws, is to lightly tap the handle-end of the screwdriver with a hammer while turning the screwdriver.

If you're thinking, "What's the big deal about removing a screw," we can assure you that a screw that's been showered on for 20 years can be pretty solidly attached to the stem. If all else fails, use an impact driver, an inexpensive tool that will remove all but the most thoroughly rusted screws. You fit the driver tip into the screwhead slot, then hit the end of the tool with a solid hammer blow. The blow seats the tip tightly in the slot and simultaneously twists the tip.

3. **Remove the escutcheon — it's probably screwed to the valve body.**
4. **Remove the bonnet nut (see Figure 11-1).**

   If this is the first time this job has been done, tile or another type of wall covering may be covering the nut. Use a cold chisel to chip away enough of the tile so you can slip a deep-set socket wrench over the nut. Deep-set socket wrenches are made just for these faucets. When you can access the nut, slip a deep-set socket wrench over the nut and turn it counterclockwise to remove the bonnet nut.
5. **Remove the stem.**

   Back the stem out of the valve body by turning it counterclockwise until it's free. The valve seat washer is at the end of the stem. The washer is held in place by a screw. Remove the screw and the washer usually comes free. If not, pry the washer off the end of the stem with a screwdriver.
6. **Take the washer to the store and purchase a replacement.**

   You have one of two basic types of washers — one is flat and the other is cone-shaped. Make sure that you purchase an exact match.
7. **Look into the valve assembly to inspect the valve seat.**

   If the valve seat looks rough or damaged, flip to Chapter 10.
8. **Replace the washer on the end of the stem and reassemble the valve.**

   Turn the stem into the valve body in a clockwise direction until it's fully seated in the valve body. Slip the bonnet nut over the stem and thread it on to the valve body. Tighten the bonnet nut hand tight, and then use a wrench to give it a full turn. Turn the water back on and open the valve. If water leaks by the stem, tighten the bonnet nut until the leaking stops.
9. **Reassemble the valve, replacing the escutcheons and handles.**

Replace the seats on both faucets, even if only one side is leaking. If one faucet is leaking, the other one will soon follow suit.

To prevent water from leaking behind the wall when you shower, use silicone or siliconized-latex caulk to fill the area between the bonnet nut and the wallcovering, or seal the back of the escutcheon.

If, when you removed the faucet, the tile or wallcovering was damaged to the point where the old escutcheons don't cover the damage, you may be able to purchase replacement escutcheons of a larger size. Otherwise, you'll need to repair the wall.

Repairing a Single-Handled Faucet

There are a great variety of single-handled faucets, and they can perform other functions besides simply selecting and mixing hot and cold water.

- They may have a built-in diverter to shuttle water between the shower or tub and may have a built-in temperature (anti-scald) control.
- The controls may go from side-to-side like a twist valve, changing the mix of hot and cold water and the pressure as they turn.
- They may move up and down or push-pull to control pressure, and twist to control temperature.

Systems vary between different manufacturers and within a single manufacturer's line.

If you're working on your shower control valve for the first time, you probably won't know exactly what you have until you take it apart. Look for the installation instructions that came with the valve — it'll be a big help. Most of these units are easy to repair and repair parts are generally available — your biggest challenge may be trying to figure out how to take them apart.

Your shower control may have a lifetime or long-term warrantee. If so, contact the company following the instructions provided with the control to obtain repair parts or a replacement shower control.

Before proceeding, turn off the water to the shower.

To access the valve, remove the handle by loosening a screw and removing the handle. You may have to tap the handle a bit with a rubber mallet to loosen it from the cartridge stem. Pull off the cap, and then remove the screws that hold the escutcheon in place. If the valve has a built-in diverter, you may have to remove the diverter before you can remove the escutcheon. A friction ring fitted between the escutcheon and the cartridge may cause some resistance in removing the escutcheon.

To take out the cartridge, either remove a clip, some retaining screws, or a cam nut. To stop leaks, you'll need to clean or replace o-rings, springs, seals or the entire cartridge.

There are three basic types of mechanisms used in shower controls: ball, ceramic disc, or cartridge. The names tend to be misleading because the ball and ceramic disc controls may well be incorporated in a cartridge. The mechanisms are far more complicated than compression faucets (see Chapter 10), and some are not meant to be disassembled. If the control malfunctions, you simply replace the cartridge. If the cartridge appears to be working okay, you may be able to replace some other parts in order to stop leaks.

Rotating ball valve

Shower valves that use a rotation ball to control the flow of water usually have a long handle that turns in a 360-degree arc. Up or down movement of the handle controls the water flow, while right or left movement controls the water temperature. This handle is held in place by a set screw at its base.

Figure 11-2 shows the parts of the rotating ball valve. These valves usually come with an adjusting tool, but it's probably long gone. Use a standard set of Allen or hex wrenches to remove this set screw. The valve mechanism is the same as for a sink faucet (see Chapter 10). The valve is usually mounted in a large trim ring. The trim ring can be removed from the valve body to expose the valve and provide limited access to the water supply pipes.

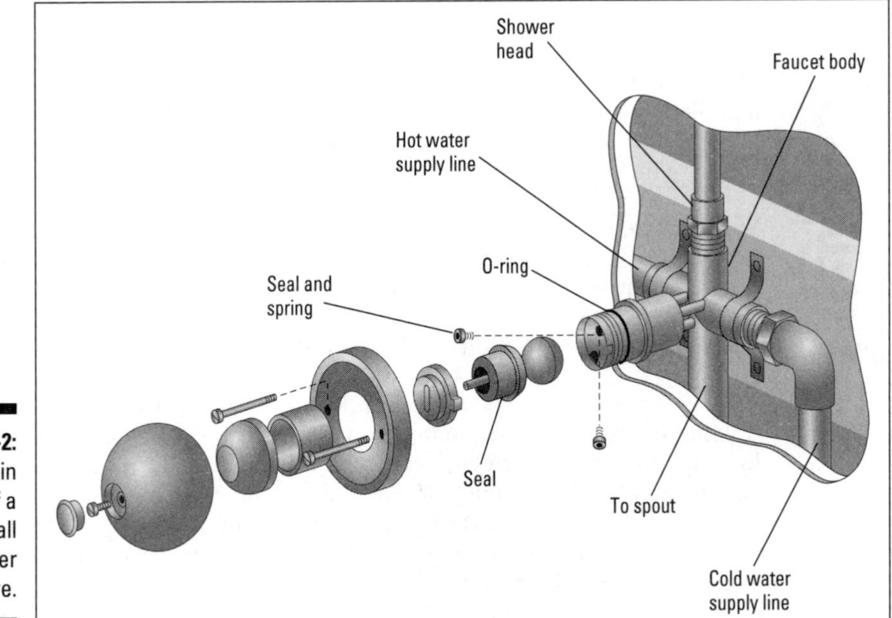

Figure 11-2: The main parts of a rotating ball shower valve.

Ceramic disk valve

The ceramic disk valve usually has a shorter handle with a more limited range of movement. This type of valve has only a few moving parts — see Figure 11-3. A set of seals is located under the cartridge, and the handle is held in place by a setscrew located in the base. Under the handle is a cover and under the cover is the disk cartridge. Try tightening the screws that hold the cartridge in place. If this doesn't stop the leak, flip to Chapter 10.

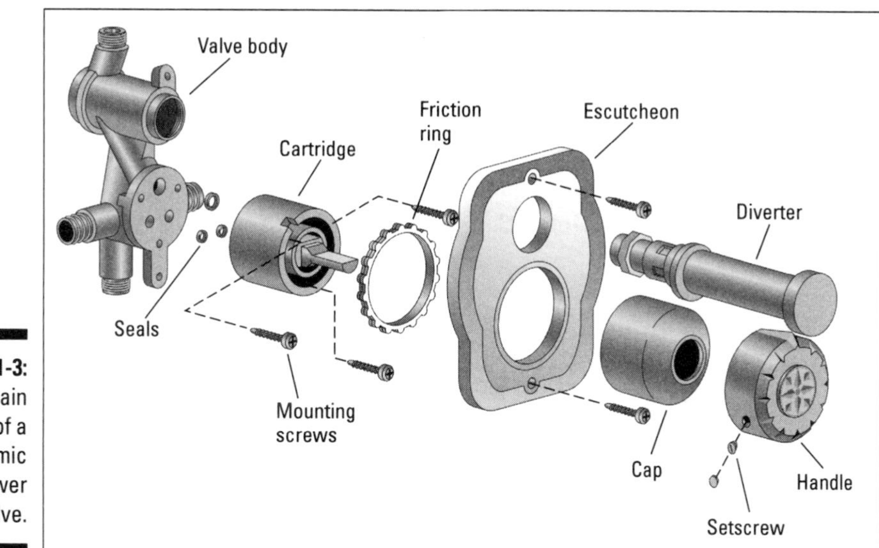

Figure 11-3: The main parts of a ceramic disk shower valve.

Cartridge valve

Tub and shower valves that contain a cartridge-type valve usually have a handle that pulls straight out to control the water flow and turns to the right or left to control the temperature. This type of valve has a large trim ring that you can remove, allowing access to the valve body and the water supply pipes leading to the valve (see Figure 11-4).

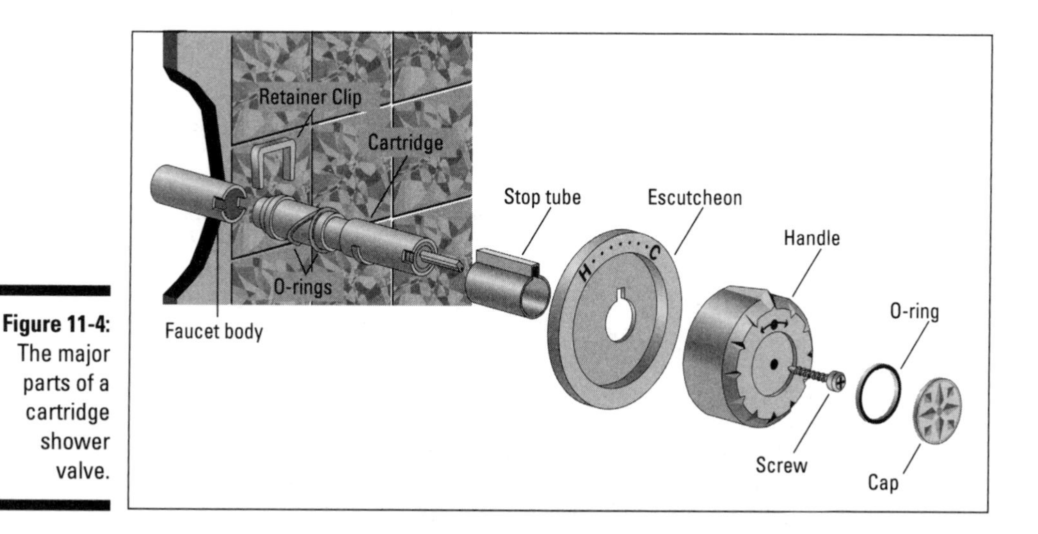

Figure 11-4: The major parts of a cartridge shower valve.

Whatever problem you encounter with this type of valve, you will most likely need to replace the cartridge. The water flow is controlled and contained by a system of o-ring seals. When these become worn, the entire cartridge assembly must be replaced.

Under a cover cap in the center of the handle is a screw that holds the handle to the cartridge stem. Remove the handle and you will see the cartridge and the U-shaped retaining clip. See Chapter 10 for help on repairing a cartridge-type faucet.

Repairing Diverter Leaks

If the tub faucet drips or dribbles water at the spout when you take a shower, the diverter valve is leaking. (This valve is usually located inside the tub spout. On some styles, the diverter is incorporated in the water control valve; on older, two-handled valves, it may be positioned between the hot and cold controls.) A *diverter valve* located in the tub spout works in conjunction with water pressure. You pull up on the diverter handle (see Figure 11-5) to close a valve that's held in the closed position by water pressure. When water to the shower in the control is turned off, the water pressure is not high enough to keep the valve closed and it drops into the open position and allows water to drain into the tub.

These valves are usually not repairable — you need to replace the spout. To remove the spout, use a pipe wrench and turn the spout counter-clockwise. Don't worry about damaging the spout — you're going to throw it in the trash anyway. If you want to try a repair, protect the finish by inserting a hammer or wrench handle into the spout to unscrew it (see Figure 11-5). If the spout doesn't want to turn, you may have to use a pipe wrench in the traditional manner. Cover the jaws with protective tape to prevent marring the spout. The best tool to use is a strap wrench, which lets you replace the spout without damage to the finish because the canvas strap grips the spout and protects it.

Replace the spout with one of equal length. The critical distance is from the base of the spout to the threaded portion. If you're unable to find an exact replacement, you need to buy a new, short piece of pipe called a *nipple* in the proper length (see Figure 11-5). When installing the nipple and spout, seal the threads with pipe dope or Teflon tape (see Chapter 3).

If the nipple is copper pipe with a threaded adapter, purchase a slip-on spout. This easy-to-install spout slides over the ½-inch copper pipe and is secured with a compression collar. You have to cut the threaded adapter off the end of the copper pipe protruding from the wall. Push the new spout in place and secure it by tightening the set screw with an Allen wrench.

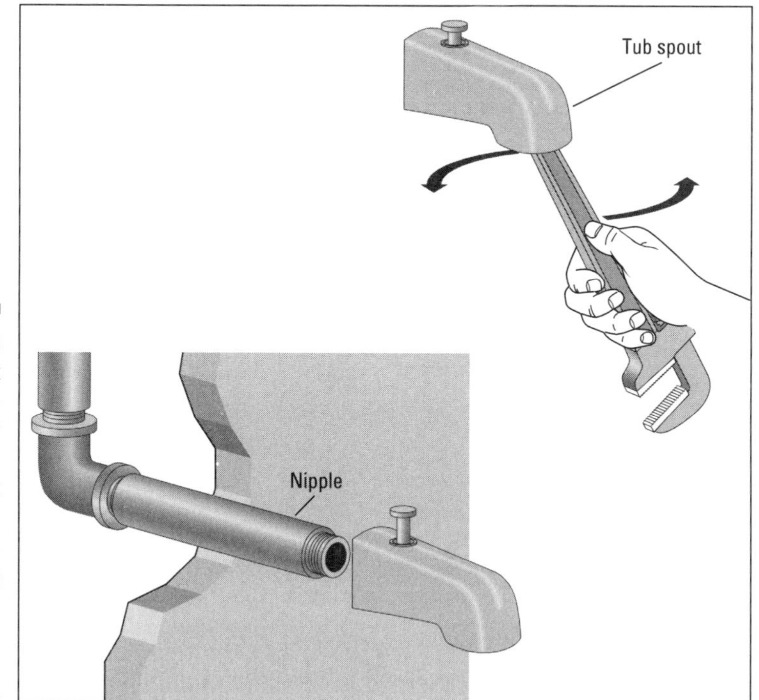

Figure 11-5: To prevent damage to the finish of the tub spout, insert a wrench handle into the spout and turn counter-clockwise.

Some tub spouts use a diverter valve built into the opening of the spout. Shut off water to the tub by pulling down on the valve, which is also held in place by water pressure. You can usually remove and replace these valves if the valve leaks water. Using pliers, unscrew the valve and inspect the seals, replacing those that are worn. Clean the valve by soaking it in vinegar to remove mineral deposits.

Fixing a Showerhead

Showerheads sometimes leak or shoot a stream of water from the threaded joint where they attach to the shower arm that extends from the wall. If the head was not tightened adequately when it was installed, constant adjustment can cause it to loosen and leak. To stop the leak, simply tighten the showerhead with an adjustable wrench (see Figure 11-6).

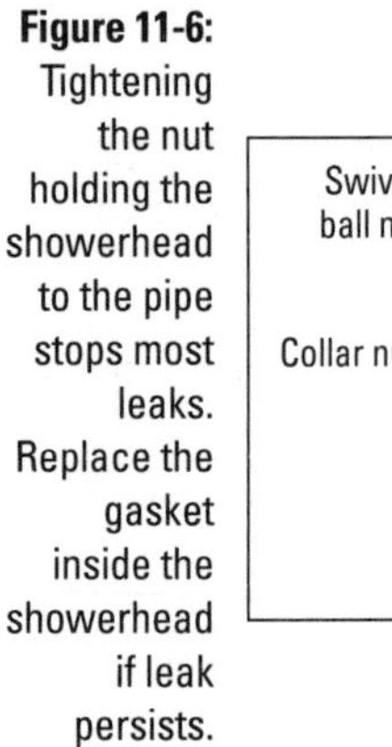

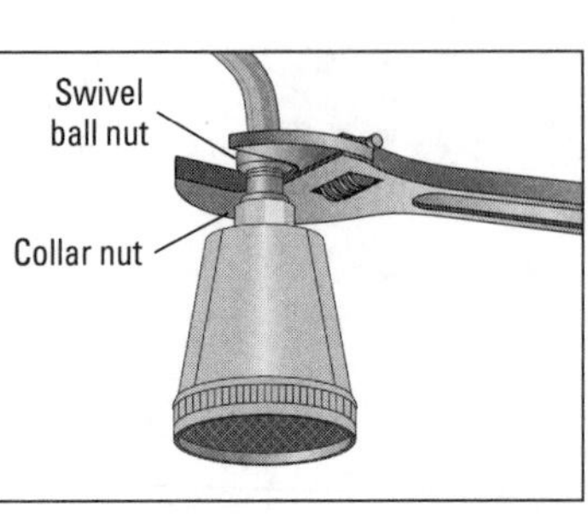

Figure 11-6: Tightening the nut holding the showerhead to the pipe stops most leaks. Replace the gasket inside the showerhead if leak persists.

If this doesn't stop the leak, remove the showerhead from the end of the pipe and look inside the threaded portion of the head that you unscrewed from the pipe. You may find a broken washer in there. Replace the washer and reinstall the showerhead.

Chapter 12

Fixing a Leaky or Run-On Toilet

In This Chapter

- Adjusting a tank handle lift chain
- Tackling a sticky tankball or flapper valve
- Fixing floatball or ballcock problems

A toilet is a pretty efficient appliance. Just push the lever and — almost magically — in a few seconds the toilet dispatches its contents with just a gurgle. Most toilets can get this job done with two or three gallons of water; newer models use only a gallon and a half. A leaking toilet, however, can waste hundreds or gallons of water a week.

If each flush does not end with a gurgle, but instead continues with a hissing sound, with water running into the toilet bowl, you have a *run-on toilet.* This is a plumbing problem that you can fix yourself. The mechanism inside the toilet bowl may seem complicated, but it really isn't.

The first thing you have to do is get up the courage to take the top off the toilet tank and familiarize yourself with the major parts — see Figure 12-1. Of course, as they do with almost every plumbing fixture, someone is always coming up with a better design. So, over time, many different types of valves and flushing mechanisms have developed, but they all accomplish the same tasks.

Here is a rundown of what happens when you push that flush lever.

1. **The flush handle lifts a round rubber *tankball* (or a rubber flapper) that's located in the base of the toilet tank.**

   When the tankball or flapper lifts, it opens the water passage between the tank and toilet bowl. As soon as this device lifts, water drains into the toilet.

2. **As the tank empties, the large ball, called a *floatball,* attached to the end of a long rod, falls with the water level in the tank.**

   Some designs have a floatball that surrounds the intake valve, which is sometimes called a *ballcock* (see Step 3).

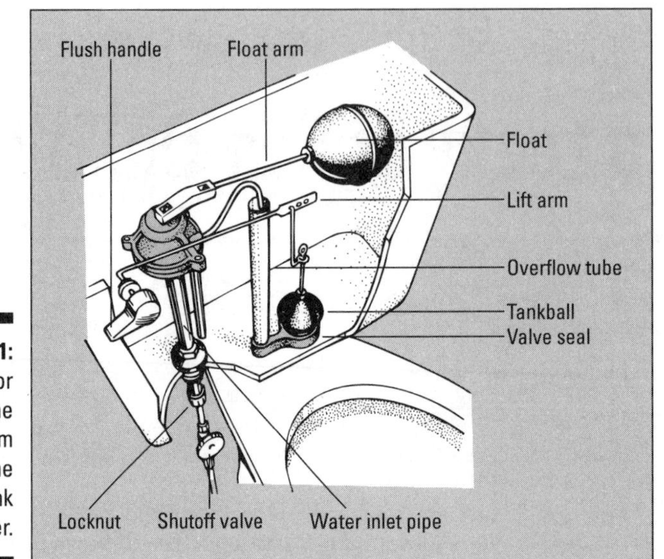

Figure 12-1: The major parts of the mechanism that fills the toilet tank with water.

3. **At the other end of the floatball rod is the *ballcock,* which opens as the floatball moves down.**

   Water begins to flow into the tank as the ballcock opens.

4. **When the tank is almost empty, the tankball or flapper falls into the outlet, stopping the flow of water out of the tank.**

5. **When the drain is closed, the tank begins to fill.**

   The ballcock also directs some water into an overflow tube that drains into the toilet bowl to assure that the bowl fill with water.

6. **As the tank fills, the floatball rises with the water level until it gets to a predetermined position and closes the ballcock, stopping the inflow of water.**

   The toilet is now ready for another flush and as long as nothing is leaking, no more water is used until the flush lever is pushed.

A run-on toilet is usually caused by a problem with the tankball, the ballcock or intake valve, or the floatball. To find the source of the trouble, remove the toilet tank top and place it in a safe location. Then push the flush lever and watch what happens.

Don't worry about the water in the toilet tank. It's clean.

Sticky Tankball or Flapper Valve

If after you flush, the water keeps running until you wiggle the flush handle up and down, the problem is probably with the linkage between the flush handle and tankball. Or you could have a bad flapper valve or tankball. The two following sections can help you fix the problem.

Fixing or replacing a tankball

The tankball is screwed on the end of a short rod that's held in place by an arm protruding from the overflow tube in the center of the toilet tank (refer to Figure 12-1). The flush lever attaches the tankball rod with another rod that slips over the end of the tankball rod. As you push the flush lever, the tankball is pulled up.

The tankball is hollow and filled with air, so as soon as it's pulled out of the drain in the bottom of the tank, it rises to the water level in the tank. But when the drain opens, the tank empties and the tankball settles back into ball seat, which is connected to the outlet pipe that leads to the toilet bowl.

To allow this open-close cycle to complete, the tankball rod and the flush lever rod need to be in alignment. If the tankball isn't falling properly into the drain, try bending the rods a bit until the tankball moves up and down without catching on anything.

To fix a sticking tankball, follow these steps:

1. **Reach into the tank and pull up on the tankball rod.**

   The rod and tankball should slide up and down easily and drop straight down into the outlet pipe. Note where the tankball hangs up.

2. **Bend the tankball rod and the flush lever rod until the tankball works freely (see Figure 12-2).**

3. **If the tankball doesn't drop directly into the outlet pipe, use a screwdriver to loosen the set screw that holds the guide arm.**

4. **Move the assembly back and forth until the tankball falls directly into the outlet pipe.**

   Check your work by flushing the toilet and making sure that the tank refills.

5. **If the tankball falls into the outlet pipe but doesn't completely stop the water flow, you may have to replace the tankball.**

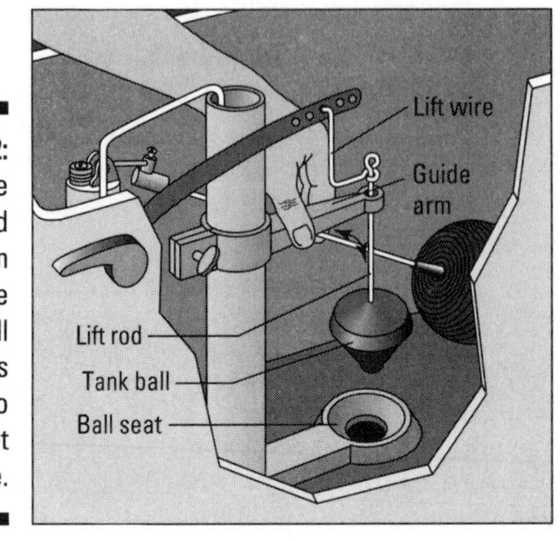

Figure 12-2: Adjust the lift rod and guide arm so the tankball drops directly into the outlet pipe.

To replace the tankball, do the following:

1. **Check that the ball seat opening (which the flapper valve falls into) is clean.**

   If you see deposits on the seat, clean it with fine steel wool.

2. **Unscrew the tankball from the end of the tankball rod (see Figure 12-2) and get a replacement.**

   Take a close look at the tankball and flush lever rods. These parts can corrode over time. These parts are inexpensive, so take the old tankball and rods with you to the store and get replacement parts that match.

If you can't get the tankball to fall into the outlet pipe and stop the water flow, buy a flapper-type tankball. Remove the old tankball by unscrewing it from the end of the brass rod (see Figure 12-3). Install the replacement by following the manufacturer's directions.

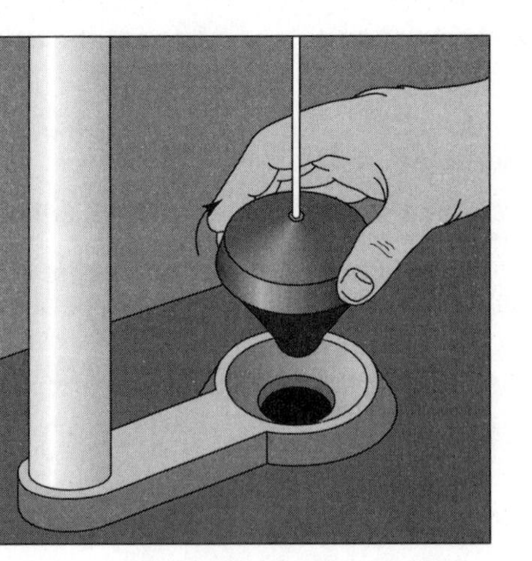

Figure 12-3: Replace a defective tankball by turning it clockwise as you hold the lift rod.

Adjusting or replacing a flapper valve

Some toilets have a flapper valve instead of a tankball. The flapper valve works the same as the tankball but it's attached to the base of the overflow tube. A hinge allows the flapper to move up and down like a door. A chain is connected to the flapper and leads up to the flush lever.

Pushing the flush lever pulls on the chain and raises the flapper valve. This action allows the water to drain out. Like the tankball, the flapper is hollow and full of air so it floats and stays in the raised, open position until the water level in the tank falls.

Here's how to adjust or replace the flapper valve:

- **Make sure that the chain that connects the flush lever with the flapper has about ½ inch of slack in it.** When you push the lever, the flapper should rise high enough to stay open but the chain should not be too tight to prevent the flapper from falling back in place.
- **Check that the flapper moves freely up and down.** To test this, reach into the tank and lift the flapper by the chain and let go. It should fall back into the outlet pipe. To adjust, loosen the screw holding the bracket around the overflow tube and move the flapper up or down, right or left as needed to establish good alignment with the outlet pipe.
- **Align the flapper with the outlet pipe.** If the flapper isn't aligned with the outlet pipe, loosen the clamp that holds the flapper in place and realign it so it falls directly into the outlet pipe. Then retighten the clamp.
- **Replace the flapper.** If the flapper falls into the outlet pipe but still doesn't completely stop the water flow, replace it. Loosen the screw that holds the bracket on the overflow pipe and pull the flapper assembly up and off the overflow pipe — see Figure 12-4. Take it to your local home center and purchase a replacement. Reinstall the new flapper in the reverse order that you removed the old, following the installation instructions provided by the manufacturer.

Floatball or Ballcock Problems

A ballcock that doesn't completely close can be another cause of a leaking toilet. A misadjusted or damaged floatball, on the other hand, is usually the cause of water dribbling into the toilet tank, running out the overflow tube into the toilet bowl, and then going down the drain.

To determine if your problem is with the floatball or ballcock, just look into the tank and note if the floatball is actually floating. If the floatball is partially submerged, it should be replaced.

If the floatball is floating, reach into the tank and lift it up. The water should stop. If it does, follow the floatball adjustment instructions in the following section. If the water continues to flow even though you're pulling up on the floatball, the problem is in the ballcock valve.

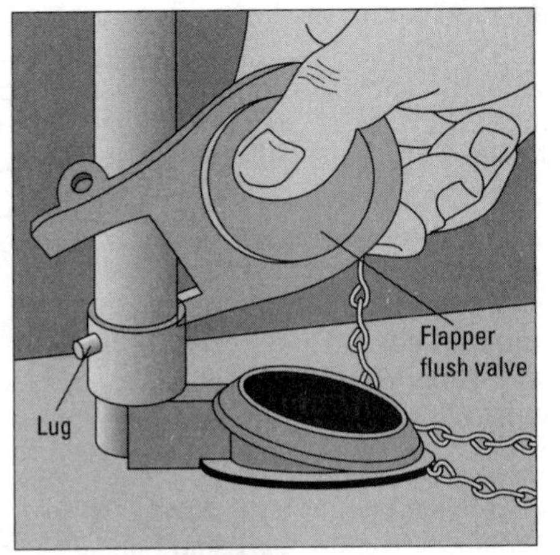

Figure 12-4: The new flapper valve slips over two lugs on the side of the overflow pipe.

Solving floatball problems

The floatball must, well, *float* in the water and resist any pressure to push it under. If it's partly full of water, replace it. Here's how to replace and adjust the floatball:

1. **Unscrew the damaged floatball from the rod by turning it counter-clockwise.**
2. **Take it to your home center or hardware store for a plastic or copper replacement floatball.**
3. **Replace the floatball by threading it on the end of the rod.**

   Turn it in a clockwise direction as you tighten it.
4. **If the water in the tank continues to run but the floatball is floating, lift up on the ball until the water stops.**

   Note the position and bend the rod down, slightly lowering the floatball and creating more pressure to close the valve as the water rises (see Figure 12-5). After bending the arm slightly, release the ball and check for running water. Repeat the process, increasing the bend in the arm, until the flow stops. Flush the toilet and check for leaking.

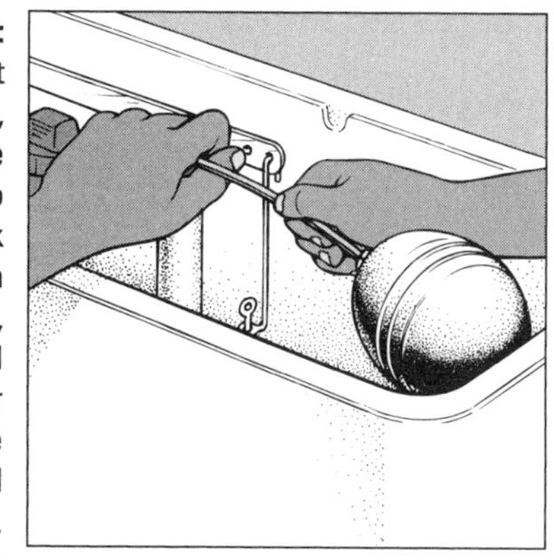

Figure 12-5: To adjust the floatball, grasp the rod close to the ballcock valve with one hand, and bend the other end with the floatball down.

Some toilets don't have a floatball on a rod, but instead have a floatball that surrounds the fill pipe. To adjust this type of floatball, loosen the screw on the side of the floatball and lower the floatball a bit on the connecting rod that leads from the floatball to the ballcock arm (see Figure 12-6).

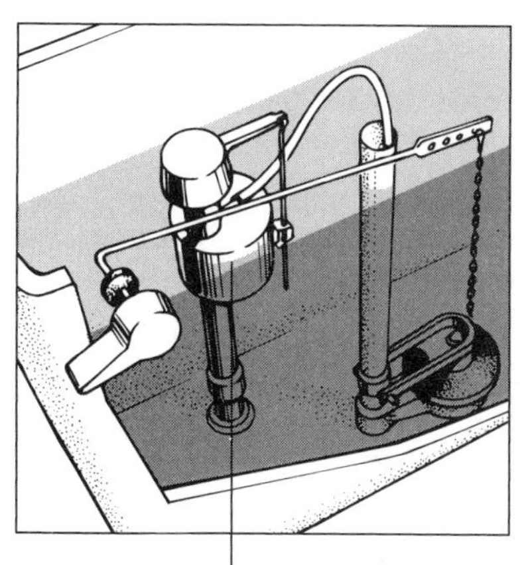

Figure 12-6: To adjust a floating cup ballcock type valve, loosen the set screw on the side of the floatball and adjust the position on the connecting rod.

If you bend the floatball arm several times and the toilet still leaks, the problem is in the ballcock. Leave ballcock repair to your plumber; ballcocks vary too much for you to quickly diagnose and repair the problem.

Addressing ballcock problems

A ballcock that keeps on leaking even after you adjust the floatball probably has some sediment in the valve body or is just worn out. Fixing a clogged ballcock isn't a project that you should undertake. With many designs that all require special parts, you may spend the rest of you life in a plumbing supply store. A plumber can tackle this job and probably has or can get the parts, but he or she will most likely recommend that you replace the ballcock.

Now, swapping out an old ballcock for a new one is definitely a project that you can tackle.

Most plumbing departments have a selection of toilet repair and rebuild kits to choose from. All come with complete installation instructions. Here are the general steps needed to replace a ballcock:

1. **Turn the water off below the toilet.**

   Flush the unit to drain the tank and then sponge out the remaining water from the bottom of the tank.

2. **Loosen the nuts securing the riser tube and remove it.**

   The *riser tube* is located under the toilet tank. This tube leads from the angle stop coming out of the wall up to the base of the ballcock valve coming out of the bottom of the toilet tank. Be sure to turn off the water before you try to remove this tube or you're going to get wet.

3. **Loosen the set screw that holds the floatball rod in place on the top of the ballcock valve. Remove the floatball rod from the ballcock assembly and set it aside.**

4. **Loosen the large nut on the underside of the bottom of the toilet tank that holds the ballcock assembly in place.**

   You may have to have a helper hold the ballcock inside the tank to keep it from turning as you loosen this nut.

5. **Pull the ballcock out of the toilet and take it with you to a home center and purchase a replacement.**

   Also purchase a new flexible plastic riser tube, just in case the distance between the bottom of the new ballcock assembly and the stop valve has changed.

6. **Insert the replacement ballcock in the opening in the bottom of the toilet tank. Thread on the retaining nut from underneath the tank. (See Figure 12-7.)**

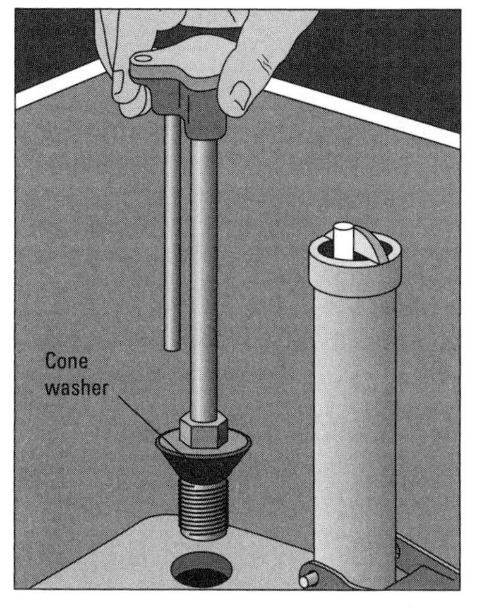

Figure 12-7: Insert the ballcock into the tank, and then thread on the retaining nut from the underside of the tank.

Follow the manufacturer's installation instructions. Don't over tighten the plastic units.

7. **Reinstall the riser tube between the stop valve and the new ballcock.**
8. **Reinstall the floatball arm if the ballcock that you purchased requires one.**
9. **Turn on the water and adjust the floatball so that the water fills to about ½ inch below the top of the overflow pipe.**

   This pipe is in the center of the toilet tank and has the flapper valve attached to it. If the water level in the tank rises too high, the excess water runs out of the tube and into the toilet bowl.
10. **Flush the toilet to test your handiwork.**

    You may have to make some slight adjustments to get the tank to fill.

Part IV
Tackling Plumbing Projects

"Sometimes Bill working for the city comes in real handy. Like when we decided to replace the kitchen fixtures."

In this part . . .

This part separates the doers from the super-doers. As a doer, you get involved in plumbing repairs and maintenance jobs just because you want to keep your water system working. You know who you are: you'll crack out a hose repair clamp kit or unclog a stuck drain with the best of them, but your heart isn't really in it.

Super-doers, on the other hand, go around the house just looking for a sink to install or a hot water heater to replace. And that's what this part is about — plumbing projects to keep you happy.

Chapter 13

Replacing a Kitchen Sink

In This Chapter

- Removing your old sink
- Knowing what to look for in a new sink
- Install your new sink and trap

Where in your house is another sink more used and abused than in your kitchen? Goodness knows what has gone down its tireless drain and been scraped off its sturdy sides, but what recognition does it get? Nada. Most people are guilty of ignoring it until, one day, a spotlight shines upon its scratched or damaged surface and suddenly its flaws are too glaring to ignore. When yours gets to that stage, this chapter can help you replace it.

Before You Begin

Most kitchen sinks are mounted in a base cabinet, and the sinks are relatively easy to replace. All you need is a screwdriver and wrenches to disconnect the water connections to the faucet and the drain lines. If you have a garbage disposer, you may find the job easier if you remove it before you remove the sink, so that it's not in the way (see the "Removing the garbage disposer" sidebar).

If you're not replacing the countertop along with the sink, the new sink has to fit the hole in the existing counter. You need to make certain the sink will fit in the original opening or make a larger opening for a larger sink. If you choose a sink that's smaller than the existing one, you'll be forced to replace the countertop because of the size difference.

If you decide to replace the sink with a larger one, be sure to measure the distance between the sides of the cabinet to ensure that you have sufficient room for the larger opening. Also make sure that internal partitions for drawers don't interfere with a larger sink.

Finding a new sink is easy, but choosing from all that's available may be more difficult. Home centers have a large selection of sinks made from a variety of materials — steel (enameled or stainless), cast iron, and composite (acrylic) materials. (All of these are described in the "Materials" section, later in this chapter.) If you decide to install a cast iron sink, you need someone to help you install it, simply because of the weight of the sink.

Removing the Old Sink

As with most plumbing projects, the first step is to turn off the water. If you have shutoff valves on the supply lines going to the faucet, turn them off. If no shutoff valves are installed, install them while you're inside the cabinet — in fact, this should be the initial job of your sink replacement. If you choose not to install shutoff valves, you must cut off the water to the whole house while you're working on the sink — the folks at home won't be happy.

Clear out all of the paraphernalia from under the sink before you begin work on the plumbing. When the area is clear, get a small bucket to catch the water in the trap. You're going to remove all of the drain lines (using slip-joint pliers), because you don't want to use them again — they're going to be filthy dirty and aren't not worth reusing. Replacing them is an inexpensive part of the project.

Removing the garbage disposer

To remove the garbage disposer, unplug it from the receptacle. If it's hard wired, turn off the electricity, and then disconnect the wiring.

The disposer will be held in place in one of three ways:

- **With a metal retaining ring:** Slide a screwdriver into one of the tabs, lift the disposer slightly, and then turn the ring counterclockwise. The unit will come loose.
- **With a stainless steel band or with two metal mounting plates bolted together:** Loosen the screw or bolts, and the disposer drops free.

The rest of the fittings and accessories are easier to remove after you get the sink out of the countertop.

Removing a self-rimming sink

When a sink is *self-rimming,* it's held in place by a built-in rim and a series of clips underneath the sink that secure the sink to the countertop. Use a screwdriver to loosen the clips and remove them. Then push up on the sink from underneath to break the seal between the sink and countertop. If it's stubborn, insert a putty knife between the rim of the sink and the countertop, and pry the sink loose.

If the self-rimming sink is cast iron, the sheer weight of the sink holds it in place and allows the sealant to firm up.

Removing a flush-mount sink

Some sinks are called *rimless* or *flush-mount* and are held in place by a separate sink rim. This system also uses clips that not only hold the rim to the countertop, but also secure the sink to the mounting rim. Because the sink and rim are separate pieces, you usually find a series of tabs that keep the sink from falling through as you remove the clips. Before you loosen the clips on a rimless sink, make sure that the tabs are bent or that corner clips are installed.

Of course, there's always the possibility the sink was installed improperly. The tabs may not have been bent or the corner clips were not installed. A heavy, flush-mounted cast iron sink may break loose because of old rusted or corroded clips. If you're removing a heavy cast iron sink, follow these safety procedures to prevent the sink from falling on you as you remove the rim clips:

- Place a 2x4 across the top of the sink that's long enough to rest on the countertop on both sides of the sink.
- Place a couple of short pieces of 1x2s or 2x2s that are long enough to span the drain holes under the sink.
- Use a short piece of rope or wire to tie the long 2x4 to the short boards under the sink.

With a nut driver or screwdriver, loosen the screws in the clips and begin removing them. Recheck that the sink is secure before you remove the last few clips. Then, lift the sink out of the opening. If it's too heavy to lift, as a cast iron sink may be, carefully lower it to the bottom of the cabinet or have someone help you lift it so that you can remove it from the kitchen.

Purchasing a Sink and a Trap

Making a decision on which sink to buy may be the toughest part of the sink-replacing process. You have a choice of materials, rimming, traps, and other options. Home centers carry a large selection, but they don't carry everything that's available. For more choices, visit a retail plumbing store to see everything that's available.

Materials

Kitchen sinks are made from steel, cast iron, or composite materials.

- **Enameled steel:** Also called *porcelain steel,* these sinks are the least expensive type. They come in many colors, but they don't hold up well. They are easy to damage with heavy pots and pans.
- **Stainless steel:** This is one of the most popular sink materials in use today because stainless steel sinks have several advantages: they're lightweight, easy-to-install, a cinch to keep clean, and they don't chip. Polishing is simply a matter of using a sponge and an abrasive cleanser.

  You can buy a stainless sink for less than $100 or more than $300. In general, you get what you pay for. The cheapest sinks are made of thinner, poorer-quality steel. They sound tinny, are easy to dent, and can be difficult to keep clean. Better-quality sinks have a sound-deadening material glued to the bottom of the sink. They're strong and remain more lustrous.
- **Cast iron:** You can be sure that an enameled cast iron sink won't break. The enamel gleams and glistens when it's new. But over time, it tends to lose its luster. You can't use a gritty cleanser, and if you drop a heavy pan into the sink you may get a chip that isn't easy to repair. But if you use care, these sinks hold up well. Cast iron sinks are available in many designs and colors. Expect to pay a premium price for these heavyweights.
- **Composite:** Sinks made from various man-made, synthetic materials have become very popular (Corian is one brand name). These materials are long-lasting and nearly impervious to damage. Scratch or burn the surface and you simply sand out the flaw. The color is the same throughout the material. Color selection is nearly unlimited. Matching countertops and sinks are available. Installation is easy, but these sinks are a bit heavier than stainless steel.

  Prices vary a great deal, but you can expect to pay more for a composite sink or one made from cast iron than for a stainless steel one.

Rimming

Some cast iron and composite sinks are self-rimming. They simply rest on top of the countertop. The relatively high rims of these sinks are attractive to some people, but the rims make cleaning the countertop more difficult because they're slightly higher than the surface of the countertop, and crumbs can accumulate there. All stainless steel sinks are self-rimming, but the rim is thinner and cleaning is not as much of a problem.

Flush-mount sinks use a metal rim to hold the sink in the countertop and seal the gap (see Figure 13-1). Food particles are easy to sweep from the counter into the sink. Some cast iron and composite sinks, along with most enameled steel sinks, are flush-mount.

Accessories and options

Look for places (additional holes) in your new sink for accessories, such as a soap or instant hot-water dispensers.

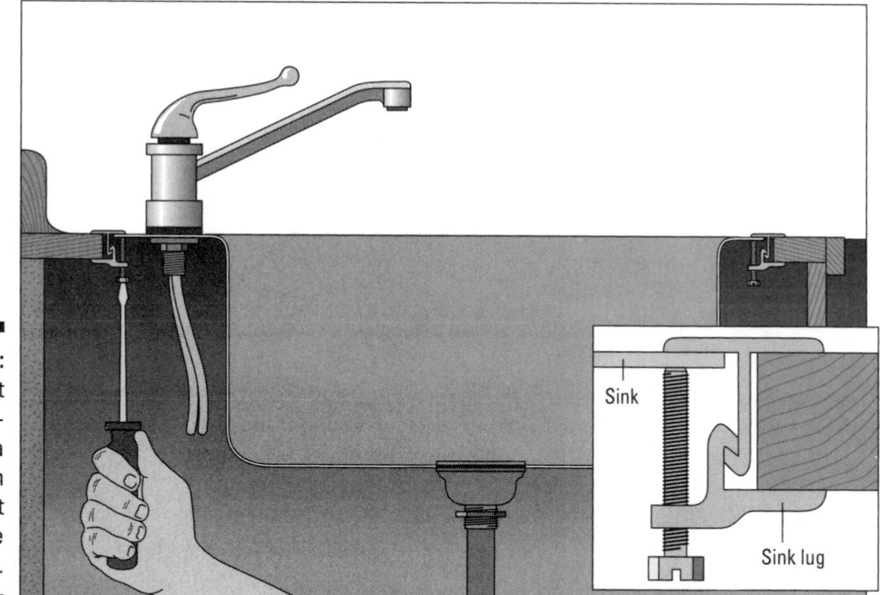

Figure 13-1: Flush-mount sinks require a separate rim to support them in the countertop.

You can drill extra holes in stainless steel and composite sinks with a metal-cutting hole saw, if you want to add these accessories later. Be certain that you have room for any accessories in the base cabinet before you cut the holes. While it's possible to cut holes in cast iron and enameled steel sinks, it's not a do-it-yourself job — it takes a professional with the proper tools to drill the hole without excessive chipping of the enamel.

Traps

Purchase the following with your new sink:

- **A waste kit with the fittings that you need for your sink configuration:** Kitchen sinks use 1½-inch diameter traps. If you're switching from a single sink to a double sink, you must connect two drains to the trap, using a tee (see Chapter 3). You can accommodate a slight change of the position of the new plumbing by using flexible drain pipes. Because kitchen sink pipes are usually not exposed, plastic lines are your best choice. Shiny, chrome-plated drain lines are pretty, but they cost more and eventually rust.
- **A waste tee with a connection for the dishwasher (if you're using a dishwasher):** Depending on the plumbing codes where you live, you may also have to install a *vacuum breaker,* which prevents the dirty water left in the dishwasher from siphoning back into your water supply system — ask the building department in your area. The garbage disposer also provides a connection for the dishwasher waste.
- **New faucet strainers:** Here's a place where saving money is usually a mistake. Cheap means leaks. Get a strainer that has machined threads where the lock nut attaches. If you're installing a garbage disposer, don't buy a strainer for that opening because the disposer has its own mounting assembly.

Puttin' in the New Sink

To work on the sink, lay it on top of the counter — you'll be more comfortable and have an easier time working with the sink. (The alternative is to work inside the cabinet, where you have to get down on your hands and knees and work in a dark hole.) Place a towel or a piece of cardboard under the sink to protect the countertop from being scratched. Attach as many accessories to the sink as you can before placing the sink in its hole in the countertop.

Installing the faucet

If you're going to reuse the faucet from the old sink, remove it and clean any caulk from the base of the faucet. Make certain the faucet isn't corroded beyond repair. This might be a good time to consider a faucet replacement/upgrade.

Install your faucet — whether it's new or from your old sink — in the new sink (see Chapter 15). To seal the joint between the faucet and sink, most use a bed of plumber's putty — similar to modeling clay, as discussed in Chapter 3, or silicone caulk. Shape the plumber's putty into a thin rope by rolling a ball between your hands. Apply the putty rope or caulk to the base of the faucet before you place it in the sink, and then tighten the mounting bolts or nuts to secure the faucet.

Plumber's putty may eventually dry, crack, or leak. Silicone caulk does a better job because it doesn't dry out — if you use clear silicone, it actually looks neater. You just have to work a little faster with silicone than with plumber's putty, however. When using it around your faucet and sink, let the excess caulk that has squeezed-out cure, before trimming it with a sharp knife. Then peel it off the sink.

Installing the strainer

A *strainer* is a cup-like metal basket that fits into the drain opening. You can find several different types of strainers, each of which has different mounting hardware. Some have a large locknut that's threaded onto the strainer bowl; others that are designed to hold garbage disposers have a flange that locks on to the base of the strainer and is held in place by screws. Install the strainer after the faucet.

Use the following steps to install a strainer:

1. **Separate all of the parts of the strainer assembly, so you can identify them.**
2. **Apply a ½-inch diameter roll of plumber's putty or a bead of silicone caulk to the underside of the flat portion of the top (called the strainer *flange*).**
3. **From inside the sink, press the strainer into the sink opening.**
4. **From the underside of the sink, press the rubber gasket onto the strainer followed by the paper gasket or friction gasket. — see Figure 13-2.**
5. **Thread the lock nut onto the strainer and tighten with slip-joint pliers (see Chapter 3), as shown in Figure 13-2.**

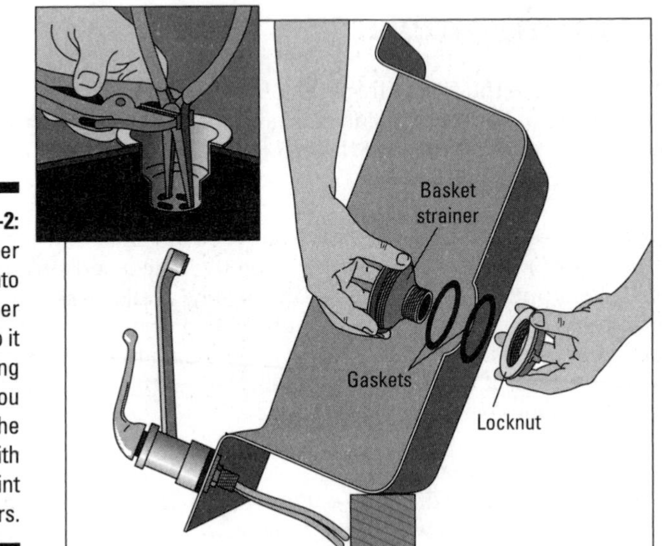

Figure 13-2: Insert plier handles into the strainer to keep it from turning while you tighten the locknut with a slip-joint pliers.

6. **Trim away the excess putty or caulk by running the edge of a screwdriver blade around the strainer flange.**
7. **Tighten the nut a bit more until there is about 1⁄32 inch of putty between the sink and the strainer flange.**

   Don't overtighten this nut or you may squeeze out all the putty.

Install the garbage disposer drain assembly by following the manufacturer's instructions, but don't install the disposer until after the sink is installed.

Installing the tailpiece

Attach the flanged *tailpiece* (the first pipe section connected to the bottom of the strainer). The tailpiece usually comes with the strainer and has a nylon insert washer the goes between the tailpiece and the bottom of the strainer.

If you're using a double-bowl sink *without* a disposer, join the two strainers with a *sink waste kit* that attaches the two strainers to one tailpiece (see Figure 13-3). You make the connection to the p-trap after the sink is installed. The *tail* and the *p-trap* are the p-shaped section of pipe at the bottom of the strainer.

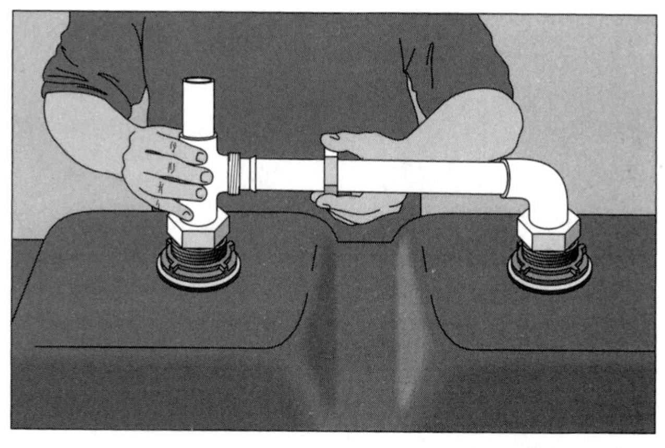

Figure 13-3: The tailpiece connects to the strainer with a slip-nut. Tighten this nut with your hand, and then turn the nut ¼ with your pliers.

If you're installing a disposer, you need to trim the connecting pipe between the disposer and the other sink. Wait until the sink is installed and you're sure where everything is located before you cut this pipe.

Installing the sink

You install a self-rimming sink differently than a flush-mount sink — follow the instructions in whichever of the following two sections that match your sink.

Installing a self-rimming sink

A self-rimmed sink rests on top of the countertop — see the "Removing a self-rimming sink" and "Rimming" sections, earlier in this chapter. If self-rimming is your preferred choice, here's what's involved to install one.

1. **Drop the sink into the cabinet opening and position it where you want it.**
2. **If you plan to use caulk instead of plumber's putty to bed the sink in, apply a strip of masking tape to the counter around the sink.**

   This makes cleanup of the excess caulk easier.
3. **Remove the sink.**
4. **Apply a bead of silicone caulk or plumber's putty along the lower edge of the rim, and then return the sink back into the opening — see Figure 13-4.**

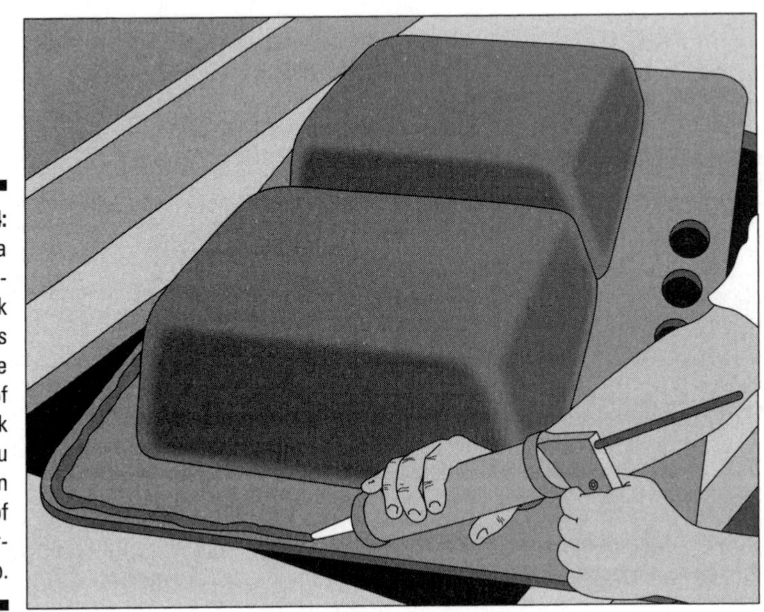

Figure 13-4: Apply a bead or silicone caulk or plumber's putty to the underside of the sink before you place it in the cutout of the countertop.

5. **Install the mounting clips to the underside of the sink and tighten the screws to pull the sink tight against the countertop.**
6. **If you applied masking tape to the counter top, remove it after you secure the sink but before the caulk sets — usually 10 to 15 minutes.**

Installing a flush-mount sink

A flush-mount sink is held in place with a metal T-shaped rim that's shaped to fit the sink. The leg of the rim fits between the sink and the countertop, while the top of the rim seals the gap between the sink and the countertop. Mounting clips hook onto a notch in the leg of the rim. As you tighten a screw on one side of the clip against the sink, it forces the sink up and the rim down, holding everything in place.

Follow these steps to install a flush-mounted sink:

1. **Attach the sink rim to the sink.**

   You need to bend a series of tabs on the rim inward to hold the rim to the sink during installation. Be sure to bend all of the tabs to assure that the sink doesn't fall through the mounting ring before you get the rim clips in place.

2. **Place the sink into the opening.**

   If you're installing a heavy cast iron sink, make sure that it's supported (see the "Removing a flush-mount sink" section, earlier in this chapter). We don't recommend working under a heavy sink that depends on only on the rim tabs to hold it up.

3. **Thread all of the clip screws into the rim clips before getting under the sink (see Figure 13-5).**

Figure 13-5: Use a nut driver or screwdriver to tighten the rim clips that hold the sink tightly to the countertop.

4. **From under the sink, attach the clips according to the manufacturer's instructions. Space them about every 8 inches.**

   Tighten the clips evenly to ensure that the corners of the sink are pulled down tight. Use all of the clips; there's no advantage in storing extra clips in a drawer.

Installing the trap

If a garbage disposer is part of the sink, reinstall it before installing the trap to the sink. This is work done under the cabinet so plan to be down on your hands and knees, hopefully with a work light hung at an angle to help you see what you're doing. Figure 13-6 can help you better understand some tricky jargon.

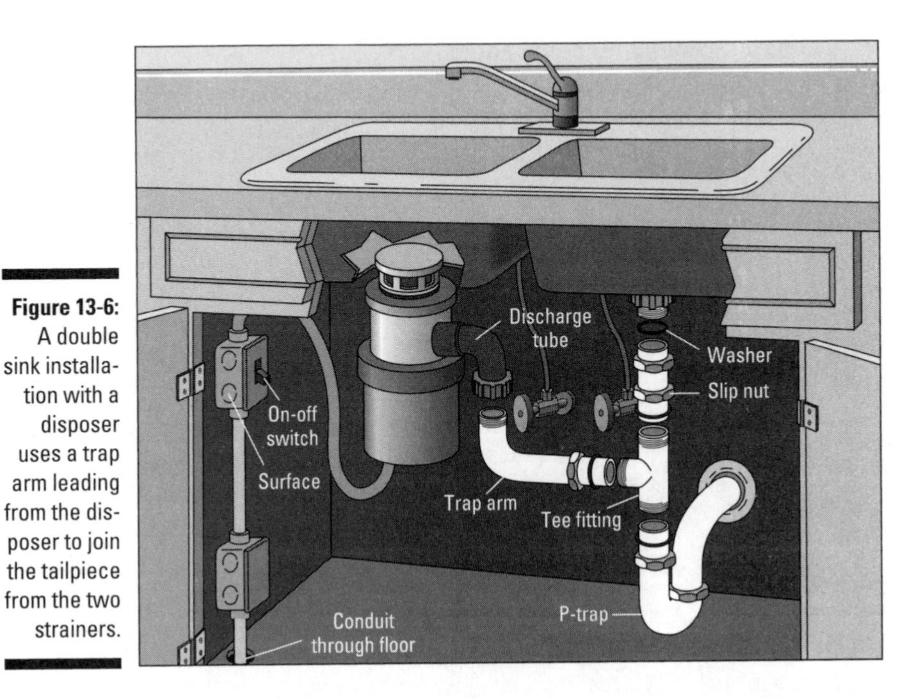

Figure 13-6: A double sink installation with a disposer uses a trap arm leading from the disposer to join the tailpiece from the two strainers.

1. **Attach a trap arm to the disposer. If you aren't installing a disposer, attach the trap arm to the other tailpiece of the second sink (refer to Figure 13-3).**
2. **Attach a trap arm to the tee fitting.**
3. **Attach the lower part of the p-trap to the tee fitting.**

   Take the p-trap apart by loosening the slip nut joining the two halves.
4. **Connect the p-trap to the drain.**

   You may have to shorten the upper part of the p-trap that leads to the drain. Test the fit by pushing the lower half of the p-trap into the drain as far as it can go. Move the upper half of the p-trap into position so the two halves mate. You can't do this if the lower part protrudes too far out of the wall — if this is the case, cut it to size with a hacksaw and try again.
5. **After all the parts are together, hand tighten all of the slip nuts and adjust all of the parts so that they are aligned.**

   Use a slip joint pliers to tighten each slip nut.

Be careful not to cross thread or misalign the threads when joining any of the fittings — especially if they're plastic. The slipnuts should tighten easily by hand. Don't force them with a wrench or pliers.

Hooking up the water

The next step is to hook up the water lines to the faucet. Use flexible plastic riser tubes (see Chapter 15), which are encased in stainless steel mesh, making them easier to install than rigid copper risers. The stainless mesh around the plastic tubing protects the tubing from damage and prevents the plastic from bursting.

Don't forget: Hot's on the left; cold's on the right.

Plug the disposer electrical cord into to the electrical outlet. Some disposers have a built-in switch, while others require an external switch. The electrical supply under the sink should be wired according to which type you have — and must meet your local electrical codes.

Testing your handiwork

Finally, you're ready to look for leaks — keep a bucket handy in the event of a failure. Turn on the water-supply valves and check for leaks in the lines that lead to the faucet. Repair any leaks before proceeding further.

Run water into the sink and check the drain lines for leaks. If a connection leaks, try tightening the nut one-quarter turn. If that doesn't work, remove the nut and spread some silicone caulk on the threads, and then retighten it. The caulk should seal the leak. Some plumbers routinely use a little caulk or plumber's putty on these joints.

Chapter 14

Installing a Pedestal Sink

In This Chapter

- Removing the existing sink
- Purchasing a new pedestal sink
- Installing your new sink

A *pedestal sink,* a sink with a bowl set on a narrow pedestal without a surrounding cabinet, can revive a bathroom. A streamlined pedestal sink takes up a fraction of the space of a boxy base cabinet. It's a particularly suitable solution to replace a damaged or dowdy wall-hung sink that no longer matches the bathroom décor. And no matter what the décor — traditional, country, or contemporary — a pedestal sink adds style and appeal.

So for whatever reason, if you're remodeling a bathroom, consider including a pedestal sink in your plan.

Before You Begin

If you have an existing *wall-hung sink* (one that's attached directly to the wall) or *vanity sink,* which is installed in a cabinet, the water and drain lines are already in place. The only special tools that you need are a *socket wrench* (which has a ratcheting mechanism to insert and remove the bolts that are shown in Figure 14-1 in a limited space) and an electric drill and bits. If you have to drill through tile to secure the sink to the floor, you need a masonry bit for your drill.

Take a close look at the existing plumbing. Older houses often have drain and water supply lines coming out of the of the floor, instead of the wall. If this is the case, these pipe will be exposed when you remove the vanity, and depending on the location, these pipes may interfere with the new pedestal sink. If you have floor-mounted pipes or you plan to install the sink at a new location (that hasn't yet had pipes run to it), you have to modify the water and drain lines. This type of work is best left to a professional — hire a plumber to do the rough-in of the new plumbing and you can install you own sink.

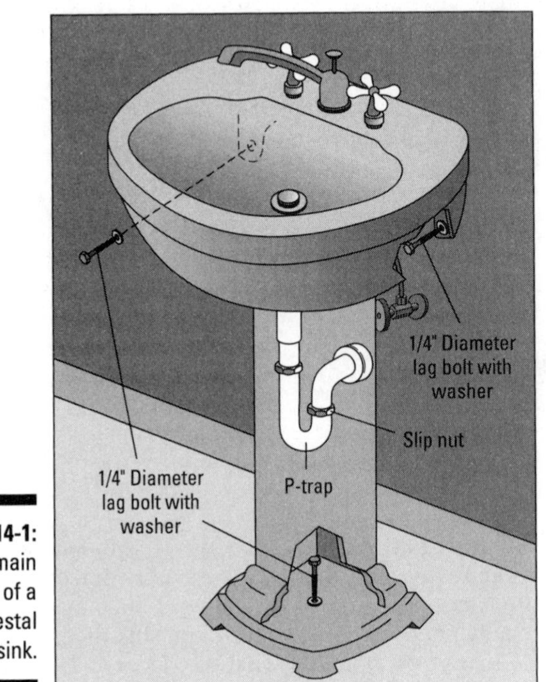

Figure 14-1: The main parts of a pedestal sink.

Removing the Old Sink

Be prepared for a sorry sight when you remove an old sink: It's likely to be an ugly patch of pocked and rough wall that was previously covered by the back of the sink. Plan to patch the holes and smooth the surface and then paint or wallpaper before installing a new pedestal sink. Install the pedestal sink only after walls are completed — as the last addition to the room.

Here's how to remove your old sink.

1. **Shut off the water supply.**

   Close the individual shut-off valves under the sink. If these valves aren't installed, shut off the main water supply valve (see Chapter 2).

2. **Open the hot and cold faucets to make sure the water pressure is released.**

   Also open faucets at the upper and lower levels in the house, if possible, to drain water from the lines.

3. **Remove the *riser tubes* (supply tubes) leading from the water supply that come out of the wall or floor to the sink valves.**

4. **Loosen the drain-pipe slip nuts with the slip-joint pliers, and remove the *p-trap*, the p-shaped section of pipe at the bottom of sink, and the *tailpiece*, the first pipe section connected to the bottom of the sink.**

   Have a bucket handy to catch the water that's in the p-trap.

5. **Remove the sink.**

   Wall-mounted sink: If you're removing a wall-mounted sink, check the two back corners of the sink to see if the sink has been screwed to the wall. If so, remove the screws. If the back has been caulked next to the wall, cut through the caulk with a utility knife. The sink should lift off of its wall bracket. See Figure 14-2.

   Remove the wall bracket and clean the wall.

Figure 14-2: After loosening the plumbing lines, lift a wall hung sink off its wall-mount bracket.

   Vanity: If you're removing a vanity, look under the sink and find the screws that hold the vanity base to the wall. Remove these screws — the vanity base and sink top should pull loose from the wall.

6. **If you plan to reuse the faucet or any of the other hardware, remove them and give them a good cleaning.**

Make A Shopping List

Before you visit a your local home center or plumbing supply showroom looking for that perfect sink, spend some time in your bathroom, assessing which size and style of pedestal sink will work best. This section gives you pointers for selecting the perfect sink.

The most important factor in determining which type of sink is the location of the plumbing in your bathroom. You can make just about any style fit by tearing the bathroom apart, but this isn't what you want nor what we recommend.

Grab your tape measure, a flashlight, a pad of paper, and a pencil and take a couple of critical measurements. Everything you need to know can be found under your present sink.

- **Measure the distance from the floor to where the drain enters the wall or floor.**
- **Measure the distance between the floor and the water-supply pipes.** If the pipes come out of the floor, consider hiring a plumber to remove them. Still, go ahead and measure the distance from the wall to the pipes.
- **Measure the distance, right and left, from each water-supply pipe to the drain.**
- **If you plan to reuse the faucet, measure the distance between the *faucet tailpieces* (pipes that protrude through the sink and are hooked up the water-supply pipes).**

You now have what plumbers call the *rough-in dimensions* for your current sink. Take these dimensions with you when you shop for a new sink. You can compare the location of your plumbing with what's required by your new sink.

Keep in mind that if your present water-supply and drain pipes are located very high on the wall, they may be difficult to mate up will a new sink. You can still use a pedestal sink, but the end result may not be as attractive.

In addition to knowing the location of your present plumbing, take a close look at the pipes and decide if they will look okay when they're exposed. Remember, the pipes will be visible and they should be clean.

- If the *escutcheons* (covers) that fit around the pipes next to the wall or floor are rusty, replace them with new ones.
- Clean corroded or rusty water-supply pipes and polish them with steel wool. Do the same with the drain pipe stub.
- If the sink doesn't have shutoff valves, install them on these and all supply lines. They make plumbing repairs easier for you and less traumatic for other residents in the household (see Chapter 2).

Purchasing a new sink

When shopping for a pedestal sink, you may have to look a little harder to find the sink you want. While these sinks are stocked by home centers and plumbing suppliers, they normally don't have a large selection on display. You may need to place a special order to get the exact sink you want.

Typically, the bowls of pedestal sinks are made from *vitreous china* (glossy porcelain enamel that's fused on metal), but you can also find them in enameled cast iron; man-made, synthetic materials; and epoxy-coated laminated wood. The material used for the pedestals may be the same as the bowl, or it can be different: for example, a wooden bowl in a metal pedestal. All of these materials should last a long time in bathroom use, but if you have reservations, check the manufacturer's warranty and find out how long the company has been in business.

Pedestal sinks may be plain and simple or have elaborate designs. The color choices are virtually unlimited.

Coordinate your sink and faucet selection. If you're reusing your current faucet, check the size before shopping for the sink. Bathroom faucets are either *fixed-width,* which means that the faucet tailpieces are a set distance apart (usually 4 inches or 8 inches), or are *adjustable* and can be adjusted up to 12 inches wide. (See Chapter 15 for the lowdown on selecting a faucet.) Lavatory sinks have holes drilled on either 4-inch or 8-inch centers. So, after selecting your sink, your choice of faucets is limited — and vice versa.

Experience has taught us that the holes in vitreous china sinks made outside of the United States aren't always accurately positioned, which may give you trouble if you're using a fixed-width faucet. If you find a sale or clearance item, have a sales clerk fit a fixed-width faucet that's similar to yours into the holes. If it doesn't fit, don't buy the sink! (If you're using an adjustable faucet, none of this matters.)

Selecting drain lines

Our favorite drain lines are plastic because they don't rust or corrode and are inexpensive — they aren't pretty, however. When the lines are exposed, shiny lines looks nicer, which makes chrome-plated brass a better choice. In fact, if you have the budget, you can special order Italian gold-plated drain lines.

Check to see if the sink comes with a drain and popup assembly. If not, you need to purchase one. Most faucet manufacturers also make matching drain and pop-up assemblies.

Buy 1¼-inch drain lines and p-trap for bathroom sinks. Compare the location of your existing plumbing with the requirements of the new sink. All manufacturers supply rough-in dimensions. This comparison will give you a good idea about whether you need to purchase an extra long tailpiece or p-trap. If the drain is very high on your wall you may even have to trim the tailpiece.

Selecting riser tubes

Use *riser tubes,* vertical water supply pipes, that's meant for faucet connections. Buy riser tubes that are slightly longer than the distance between the faucet connection and the shutoff valve.

Chrome-plated soft-copper supply tubes may be the best choice because they look nice and are reasonably easy to install. Those with flexible ribs are nearly impossible to kink. If you use a soft-copper tubing without the ribs, bend it with care or use a spring tubing bender to prevent kinking the tubing.

If the pipes are located where they're mostly out of sight, you may want to use polybutylene risers. They are less-expensive, easy-to-install, flexible, corrosion-proof riser tubes that you can cut to length with a knife. To measure, hold the tube in position, making the bends as you hold it. Mark the length, and then let it straighten for cutting.

Installing the Sink

When you're unpacking the sink, be careful not to drop any of the parts. The base and bowl are heavy and sometimes they're awkward to handle. Use caution and store them in a safe place, out of harm's way.

Use the manufacturer's installation instruction to identify the parts of the sink along with the hardware to assemble it. Lay out the pieces on the floor and refer to them as you read the instructions.

Installing the wall bracket

Consult the manufacturer's installation instruction on where to position the *wall bracket,* on which the sink rests. Most wall brackets should be installed slightly below the final height of the sink. This is an important measurement because if the sink is mounted too low, the pedestal won't fit under it; if it's too high the sink, won't rest properly on the pedestal.

Do you need a backer board?

Even though the bottom of the bowl is supported by the pedestal, a good pull on or an accidental body blow to the sink could dislodge it. If you find that you're drilling through only drywall and not into solid wood, the installation instructions provided by the manufacturer may call for a backer board. Unless the directions explicitly tell you to use wood, you can install a section of cement board as a backer. Cement board is made of an aggregated Portland cement core that's sandwiched between two layers of fiber mesh. This is much stronger than drywall and gives better support to the sink.

Many sinks require a wood backer board. To install a wood backer board, do the following:

1. **Remove enough drywall — and if you have it, tile — to make room for the board.**

   If you have tiled walls or another wall covering that you don't want to disturb, consider working on the wall from the opposite side of the bathroom wall. If it's a closet, you're in luck. Even if it's a living area, drywall repair is easier than tile repair.

2. **Install the backer board so that the wall bracket lag bolts are centered vertically in the board.**

   Fit the backer board into notches cut into the studs. If pipes prevent use of a 2x4, use a 1x6 instead. Mount the backer board to the studs with nails or screws. (See the following figure.)

3. **If you're coming in from the opposite side of the wall, cut 2x4s about 18 inches long and notch them for the backer board and nail it in place. Fasten the 2x4s to the studs so that the backer board is next to the bathroom wall. (See the figure, below.)**

4. **After the backer board is installed, repair the wall.**

   Cut a piece of drywall to fit the hole in the wall, and use drywall screws to fasten it to the wall framing. Apply a coat of drywall compound, and embed drywall tape in the compound. When dry, apply several more coats of drywall compound, allowing each coat to dry. Sand smooth and paint.

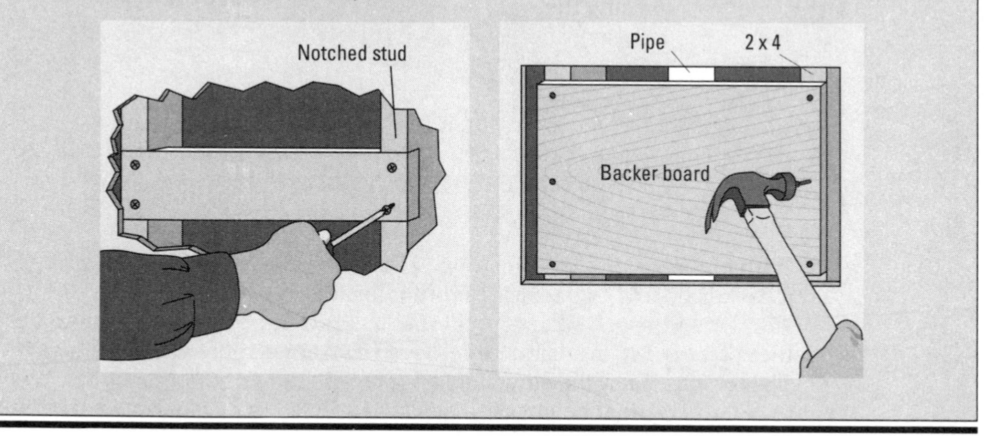

Install the wall bracket only into solid wood — not into drywall. Drywall isn't strong enough to support the sink. See the "Do you need a backer board?" sidebar for more information.

Follows these steps to mount the pedestal sink and its base.

1. **Drill pilot holes for the mounting bolts.**

   Use the wall bracket as a template if the manufacturer does not supply one: Locate the bracket over the drain pipe at the recommended height, and, using a pencil, mark the location of the mounting bolts on the wall.

2. **Fasten the bracket to the wall.**

   With toggle bolts: If you have plaster walls, mount the bracket to the wall with large *toggle bolts,* which are screw devices with collapsible wings that push up against the back side of the plaster when the fastener is tightened — see Figure 14-3. Drill pilot holes for the toggle bolts. Assemble the toggle bolts in the bracket and push the toggle bolts into the wall. The wings will spring open behind the wall and prevent the bolt from pulling out. Tighten the bolts and the bracket is pulled tight against the wall.

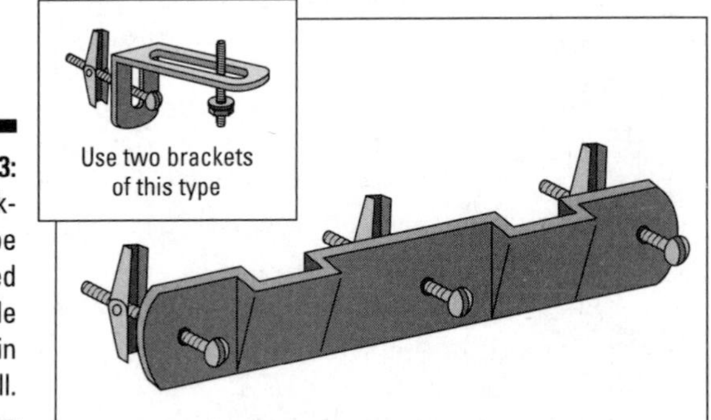

Figure 14-3: Wall brackets can be installed with toggle bolts in plaster wall.

   With lag bolts: If you've installed a backer board (see the "Do you need a backer board?" sidebar for more information), drill a pilot hole in wood about ⅛-inch smaller than the lag bolts supplied by the manufacturer. (A *lag bolt* is a large, pointed screw with a square or hex head.) Tighten the lag bolts with a socket and ratchet wrench.

   If you have tile on the walls, use a carbide-tipped drill bit that's made for drilling into tile. Drill a hole the same diameter as the lag bolt but stop short of drilling into the wood backer board. To make a good starting point for the drill bit, nick the tile by tapping it carefully with a sharp nail or punch to mark and begin the hole for the drill bit. Drill on slow speed to avoid breaking the tile.

Some pedestal sinks don't use a wall bracket. Instead, these models are mounted to the wall with lag screws or toggle bolts through holes in the back of the bowl. Other manufacturers provide a way to fasten the bowl to the pedestal. If your sink doesn't use a bracket, don't fasten it to the wall until after you install the faucet.

Installing the faucet assembly

Flip to Chapter 15 for detailed instructions on installing a faucet assembly.

Installing the drain and pop-up assembly

The drain is assembled when you buy it, so your first step is to disassemble it. Read the installation instructions in case there is something unique about the unit, but follow these basic steps to install the drain and pop-up assembly:

1. **Place a bead of plumber's putty or silicone caulk on the drain flange and insert it into the basin opening.**

   See the "Working with silicone caulk" sidebar for more information. The drain *flange* (rim around the drain) may screw onto the drain stem or the base of the drain, or it may be immovable.

2. **Slip the wedged-shaped sealing washer onto the stem from the bottom of the bowl, followed by a washer and the fastening nut.**

3. **Tighten the nut until the putty or caulk is compressed and the flange looks and feels tight.**

4. **Loosen the nut holding the pivot rod to the back of the drain stem and insert the pop-up plunger into the drain.**

Working with silicone caulk

Silicone caulk can be difficult to use, but in applications where you may have some movement along with water, it does a better job of sealing than anything else. When using it around faucets and sinks, let the caulk cure (dry) before turning on any water. When it has cured, use a knife to trim the excess caulk from around the fixture and on surrounding areas. You have a very short working time and if you make a mistake, you're better off letting the caulk cure before you finish the installation, and then removing the caulk and starting over.

If you want some of the advantages of silicone caulk, but with more working time, siliconized latex caulk is a good choice.

Yours may be one of several different types of plungers. Some can't be removed until the pivot rod is removed; others simply twist onto the lever and are easily removed.

5. **Slide the end of the pivot rod through the opening in the plunger and tighten the nut until it feels snug.**
6. **From the top of the faucet, insert the lift rod.**
7. **Pull the pivot rod all the way down.**
8. **Connect the *clevis,* which is the short arm through which the pivot rod extends.**

   Half of the spring clip should go on each side of the clevis.
9. **Place the pivot rod in the second hole of the clevis.**
10. **Slide the lift rod into the clevis and pull the rod all the way down.**
11. **Tighten the clevis screw — see Figure 14-4.**

Hanging the sink

Most pedestal sinks can hang on the wall without the pedestal. We suggest mounting the sink without the pedestal in the way — it's easier to install the risers and p-trap. To hang the sink, set the bowl on the wall bracket and pedestal. Follow the manufacturer's installation instructions.

Some sinks have additional mounting hardware. You may find a couple of holes under the sink through which you apply additional mounting screws — this holds the sink tightly against the wall and firmly on the bracket.

Making the connection

With the sink on the wall you can get under it fairly easily. All that's left to do is install the drain and water supply lines.

1. **Install the p-trap.**

   You may have to cut the tailpiece that protrudes from the pop-up assembly, if the p-trap doesn't align with the drain that protrudes from the wall. If this is the case, be sure to push the p-trap all the way up on the tailpiece. Then measure the distance from the center of the p-trap to the center of the wall drain. Cut this amount from the bottom of the tailpiece.

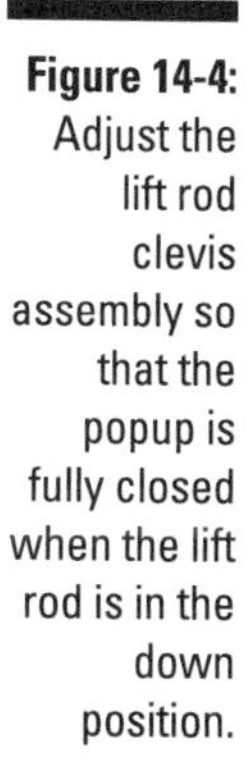

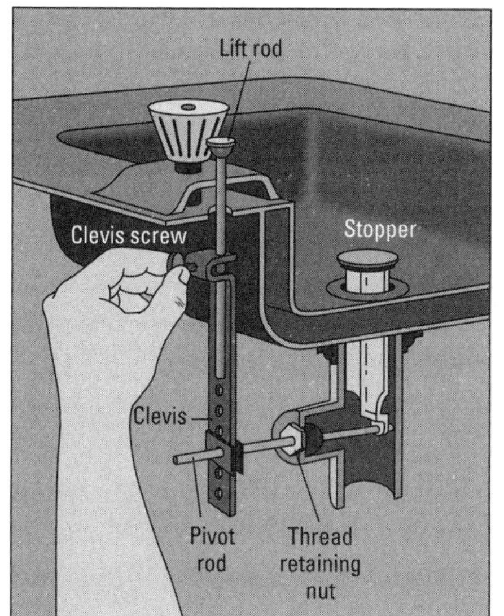

Figure 14-4: Adjust the lift rod clevis assembly so that the popup is fully closed when the lift rod is in the down position.

Take the p-trap apart and install the top part of the trap onto the tailpiece. Insert the lower part into the wall drain. Pull the lower part of the trap forward, move the upper part in alignment with the lower, and tighten the slipnut to join the trap parts together. Tighten the slipnut on the tailpiece and the slipnut on the wall drain fitting. See Figure 14-5.

Figure 14-5: Slide the upper part of the p-trap into the drain in the wall. Slip the lower part on to the sink tailpiece, then tighten slipnuts.

2. **Install the riser tubes.**

   Attach the riser tubes to the faucet tailpieces using the compression nuts supplied with the faucet. Snug up the nuts tightly, by hand.

3. **Bend the risers so that they lead to the stop valves protruding from the wall. Mark the tubes, remove them, and cut them to length.**
4. **Reinstall the riser tubes and hand tighten, and then take a turn or so with a wrench. (See Chapter 15 for complete instructions.)**

Finishing up

You're just about done, so hang in there. Go through these simple tests to make sure that you have a perfect installation.

1. **Turn on each shutoff valve and check for leaks.**
2. **Try the faucet and check for drain leaks.**
3. **Test the pop-up valve.**

   If the valve doesn't close completely, adjust the position of the clevis on the lift rod.
4. **Place the pedestal in position and make final adjustments in the position of the pedestal.**

   If there are provisions for anchoring the pedestal to the floor, do that now, as shown in Figure 14-6.

Figure 14-6: After checking for leaks, slide the pedestal under the sink and secure the pedestal to the floor with lag bolts.

5. **Caulk any gaps or voids between the wall and bowl.**

   Use a colored latex caulk that matches the color of the sink as closely as possible. Force the caulk into the voids as deeply as possible.

6. **Clean any excess caulk with a wet rag.**

Chapter 15

Installing or Replacing a Sink Faucet

In This Chapter

- Removing the old faucet
- Purchasing a new faucet
- Installing a faucet

Replacing a faucet isn't a difficult job — you don't even need special tools, because faucets come with complete installation instructions that are easy to follow. As a matter of fact, faucets are designed for do-it-yourself installation. While installing a faucet can be quite simple, consider all of the following factors before deciding whether you want to install a faucet yourself.

- **Are you claustrophobic?** You have to work in tight quarters and may be uncomfortable while you're working. A kitchen cabinet may be quite small — and a bathroom vanity cabinet even smaller than that — so if you're claustrophobic, you might want to hire a plumber.
- **How old is your house?** The age of the house and how long the existing faucet has been installed determines what kind of plumbing fittings are involved. Check the supply pipes: If they're galvanized iron and connect directly to the faucet without supply tubes (called *risers*), you have some difficult work ahead.

  Old pipes are difficult to work with. What may start out as a simple project can become a disaster if you break an old, brittle pipe that's inside the wall. If the plumbing is old but has been upgraded along the way, you can do the job.
- **Have the fastening nuts rusted?** Getting the old faucet out can be difficult if the fastening nuts are rusted or made of dissimilar metals, causing the nuts to fuse to the faucet bolts.

- **Are shutoff valves installed?** If you don't have shutoff valves between the permanent piping and the supply tubes, you should add them.
- **Are you working odd hours?** Double-check whether the plumbing department where you shop is open for business during the hours that you plan to work on the project.

If any of these are concerns of yours, call a plumber. Otherwise, installing a faucet is a straightforward job that you can complete in half a day, assuming you don't run into any complications.

Replacing a Kitchen Sink Faucet

Hooray for the hard-working faucet in the heart of the house: the kitchen sink faucet. Replacing one is a doable job that everyone in the household will appreciate because this faucet gets such a workout. We don't know of another faucet that gets manhandled by so many thirsty, dirty, hungry, and industrious hands.

Pulling out the old faucet

Begin this project by cleaning out all of the supplies and miscellaneous items stored under the sink. This in itself could take you all day! You'll also have more room if you remove the *waste drain* that takes waste water out of the house, and the *p-trap,* the p-shaped section of pipe at the bottom of the sink.

An inexpensive *basin wrench,* shown in Figure 15-1, can save your knuckles. This wrench is designed to reach up from underneath the sink to get at the hard-to-reach nuts that hold the faucet to the sink, and to loosen the supply tub nuts.

Here's how you remove your old kitchen faucet:

1. **Turn off the water supply to the sink and open the faucet to relieve the water pressure (so that you don't get a spray in the face when you loosen the riser tube nuts in Step 2).**

   We recommend installing shutoff valves on all supply lines because they make plumbing repairs easier and less traumatic for other residents in your house. Shutoff valves allow you the luxury of turning off the water supply to only the fixture that you're working on, so the other water supply in the house can continue to flow as normal. See the "Adding a shutoff valve" sidebar for tips on completing this task.

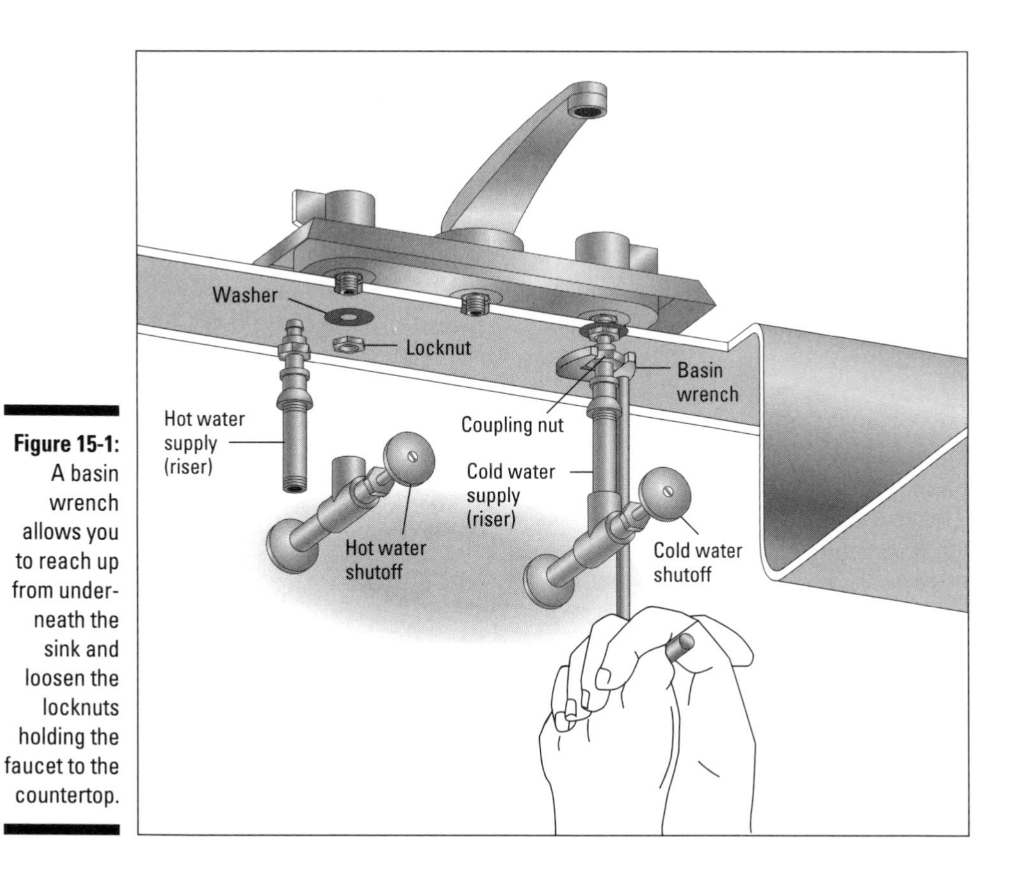

Figure 15-1: A basin wrench allows you to reach up from underneath the sink and loosen the locknuts holding the faucet to the countertop.

2. **Loosen the nuts and the lower end of the riser tubes that hold the stop valves (see the "Adding a shutoff or stop valve" sidebar).**
3. **From below the sink, use a basin wrench to loosen the nuts that attach the top of the riser tubes to the tailpieces of the faucet.**

   If the stop valve has copper tubes coming out of its body, use two wrenches; one on the fitting on the end of the copper tube and the other to loosen the riser tube nut. To remove the riser tubes, you may have to wiggle them a bit to get them loose.
4. **Using a basin wrench, remove the locknuts that hold the sink to the countertop.**

   If these nuts are rusted, apply penetrating oil to the nuts and allow them to stand for 15 or 20 minutes before you try again. If that doesn't work, use a hammer and chisel or a hacksaw to loosen nuts. Keep in mind, however, that using these tools in a confined space is difficult, at best.

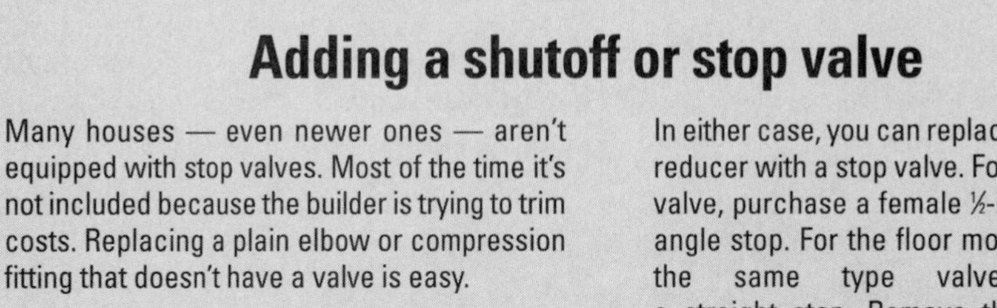

Adding a shutoff or stop valve

Many houses — even newer ones — aren't equipped with stop valves. Most of the time it's not included because the builder is trying to trim costs. Replacing a plain elbow or compression fitting that doesn't have a valve is easy.

Depending on the type of plumbing pipes that you have, you have a choice of the type of *stop valve* (which controls the water flow to the fixture) that you can install. Here's the rundown on installing a stop valve on the three type of pipes you usually encounter.

- **Galvanized pipe:** Most likely, you'll find a pipe coming out of the wall or the floor. (If this is a hard metal pipe, it's a galvanized pipe.) A pipe coming out of the wall has an elbow that turns up towards the fixture. From this elbow, a riser tube brings water to the fixture. If the pipe comes out of the floor, it most likely has a reducing fitting that joins the pipe to the riser tube leading to the fixture.

  In either case, you can replace the elbow or reducer with a stop valve. For a wall mount valve, purchase a female ½-inch by ⅜-inch angle stop. For the floor mount, purchase the same type valve, but get a straight stop. Remove the old piping, apply pipe dope or Teflon tape to the end of the pipe, and thread it on the new stop valve. Tighten with an adjustable wrench.

- **Plastic or copper pipe:** If you have plastic or copper piping, you have to cut the pipe just before the elbow with a hacksaw or tubing cutter (if you have room to work the tool). Then, install a compression-type stop valve. Slip this type of valve over the pipe and secure it in place by tightening the compression nut.

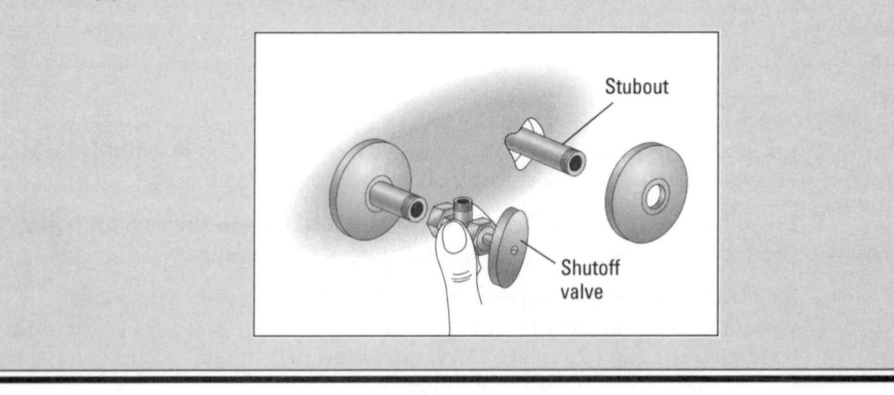

Of course, if you have to revert to the hacksaw, you're going to wreck the faucet to get it out.

5. **Faucets that have a pullout spout usually have a center mounting nut that must be removed.**

On these faucets, the spout hose loops down below the countertop. You have to remove the hose from the valve body before you can completely lift the faucet off the counter. Remove the riser tubes and the mounting bolts, and then pull the faucet partially out of the sink — you should be able to remove the hose. After the hose is off, the faucet comes completely out of the countertop.

Purchasing a new kitchen faucet

You may need more time to choose a new kitchen sink faucet than you do to actually replace it. Take a walk through the plumbing department of a home center and you see a mile-long (slight exaggeration) display of kitchen sink faucets. As kitchens have become designer showcases as well as functional cooking centers, the choice of accessories — especially with faucets — is far more abundant than it was 20 years ago. Here are some of the popular options and features to consider:

- **Number of handles:** Two-handle and single-handle faucets are the basic categories that are the most convenient.
- **Fixed or swing spouts:** A fixed spout is fine for a single-bowl sink. in which the flow of water is directed into one bowl. However, for a sink with two bowls, a swing spout is essential so it can direct water into both of the bowls.
- **Hose sprayers:** These are optional, but are convenient for spraying vegetables and cleaning the sink. They don't have to be made out of black plastic. You can find nearly as many style and color options for sprayers as you can for faucets.
- **Single-handle faucets with pullout spouts:** These faucets are becoming quite popular — you can pull the faucet out of its holder and with the flick of a switch, turn it into a sprayer. They are handy to use and will put the water where you need it, even into a pot sitting next to the sink.
- **High-reach faucets:** Faucets with high-reach spouts are no longer designed just for bar sinks. They're made in single- and two-handle models.
- **Unique designs:** Faucets that you thought were old-fashioned are now in vogue — many that look like they came out of an old house. You can also find sinks with empty holes designed for soap dispensers.
- **Body material:** Bodies made of brass, stainless steel, or plastic. Metal bodies are long lasting. Plastic faucets, however, are a good choice for a laundry room or shop sink, because they're inexpensive.

All kitchen and bathroom faucets sold must meet federal regulations intended to minimize leaching of lead. Water dissolves lead from the brass used in faucets. Brass is still used for faucet bodies, but manufacturers have found ways to prevent lead from leaching into the water.

- **Colors and finishes:** Chrome, brass, and combinations of the two are just the beginning. Solid, vibrant colors are also available.

  Finishes with lifetime warranties are guaranteed not to corrode, scratch, or mar. Save your receipt so you can prove that you bought the faucet.

- **Built-in purifiers:** These filter out lead, chlorine, and other contaminants. Some faucets have filters built into the unit; others have filters that attach to the spigot. Those with a pullout faucet wand are available with a filter built into the wand. These filters have replaceable cartridges that last about three months.

Your new faucet — no matter what the style — must fit into your existing sink. Take the old faucet with you when you go shopping. The most important feature of the new faucet is how many mounting holes it requires the sink to have and the spacing between these holes. Many faucets require a center hole and come with a wide top trim cover that will conceal the holes used by your old faucet — see Figure 15-2.

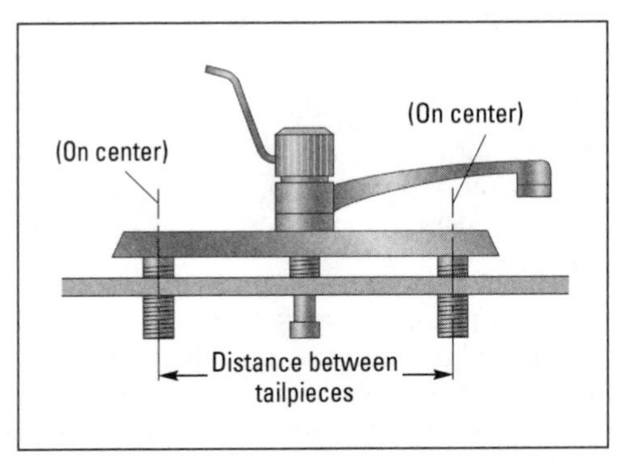

Figure 15-2: The most important factor to consider in purchasing a new faucet is the distance between the tailpieces. The faucet must fit your sink.

You can purchase a faucet with plastic bodies for as little as $50. For a quality faucet, expect to pay around $100, while a top-of-the-line faucet can cost more than $300.

Selecting a trouble-free supply (riser) tube

Making sure the water gets from the faucet to the water supply is the job of the supply tubes (also called *risers*) found under the sink. You have three choices in tubes:

- **Polymer lining and stainless steel mesh:** Our first choice for supply tubes are those made of polymer lining and stainless steel mesh. Even though they're the most expensive kind, we've never had one fail. Each end is fitted with a threaded fitting. These fittings attach directly to the faucet tailpiece and stop valve and don't rely on compression nuts. They're strong and flexible, but can't be cut to exact lengths — that's why they're available in several lengths. Buy tubes slightly longer than the distance between the faucet connection and the shutoff valve, so that they're not stretched tight. If the tube is several inches too long, just let it form a loop.
- **Soft-copper:** Plated soft-copper supply tubes look nice and are reasonably easy to install. Those with flexible ribs are nearly impossible to kink. Use risers with a formed end for faucet connections. When measuring, hold the tube against the faucet and make the necessary bends so that the tube attaches to each fitting straight on.
- **Polybutylene:** The least expensive variety of tube is polybutylene: It's flexible, corrosion-proof, and can be cut to length with a knife. To measure, hold the tube in position, making the bends as you hold it. Mark the length, and then let it straighten for cutting.

Installing the faucet

Okay, you've removed the old faucet, chosen a new one, and have read the directions from the manufacturer tucked inside the package. (If the faucet is going on a new sink, you have the sink propped up carefully on the countertop and it's at a convenient location to work on.)

If you're replacing an old faucet, remove the buildup of minerals and soap that accumulate around the old base plate. Start with soap and water, and if that doesn't work try vinegar or a cleaning product that's designed to dissolve mineral deposits. Clean the sink with a sponge and cleaning powder. If you have a tough buildup of minerals or soap scum, use a single-edge razor blade that's laid nearly flat, but be careful not to scratch the sink.

Install the new faucet by following these easy steps:

1. **Unpack the new faucet, laying out all of the parts and rereading the directions to identify the names of the parts.**
2. **If you have a hose spray, install it first.**

   Place the hose guide into the hole in the right side of the sink. Feed the hose down through the hose guide. From underneath the sink, wrap some Teflon tape around the male threads and tighten the hose onto the bottom of the faucet. (See Figure 15-3.)

   Feed the spray hose through the sink hole on the right, slide the faucet shank into the hole, and attach the fastening nut from below — see

Figure 15-3. Feed the hose up through the center hole of the sink and through the base gasket (if you have one). Wrap some Teflon tape around the male threads and tighten the hose onto the bottom of the faucet.

3. **Align the faucet tailpiece with the holes in the sink and lower it into place on the sink.**

   Some faucets use gaskets to prevent leaks in the base of the faucet, while other manufacturers recommend using plumber's putty around the base. You can also run a bead of silicone caulk around the base instead of using plumber's putty.

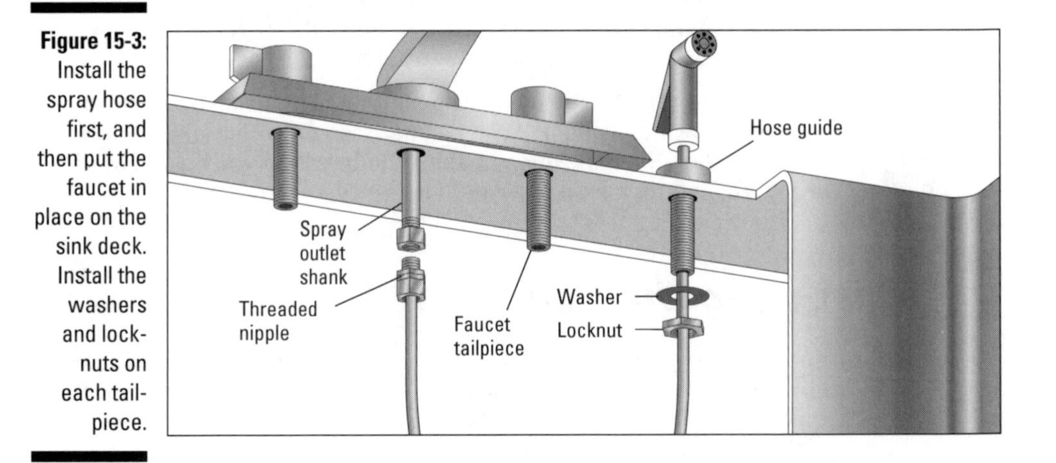

Figure 15-3: Install the spray hose first, and then put the faucet in place on the sink deck. Install the washers and locknuts on each tailpiece.

4. **From under the sink, thread the fastening nuts onto the faucet shanks and draw them up finger tight (see Figure 15-4).**

   Align the faucet base so that it's parallel with the back of the sink. Tighten the nuts with pliers, a basin wrench, or a long-reach socket. Some faucets have plastic nuts that you're supposed to be able to fasten with your fingers, but an extra turn with slip-joint pliers ensures a tight fit.

   Some single-handle faucets don't have bolts or shanks. Instead, they use a large nut threaded onto the main faucet body. The mounting principle is the same, but there's just one nut.

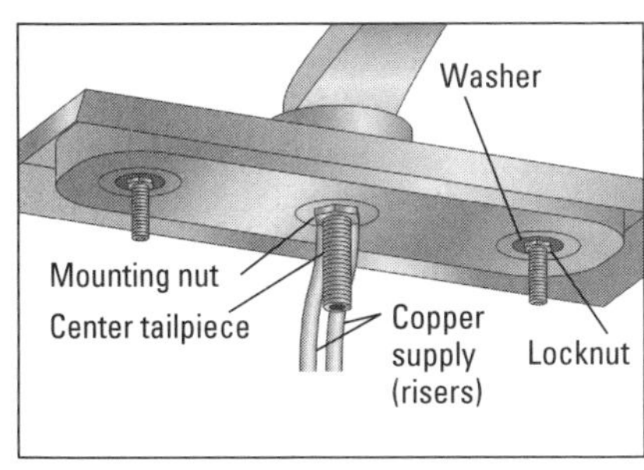

Figure 15-4: Some faucets mount with three bolts and have risers attached to the valve body.

Hooking up the water (installing the riser tubes)

With the faucet in place you can connect the faucet to the water supply. Go easy during this operation. You may have battled rusty nuts during this project, but installing riser tubes requires a light touch.

Grab the riser tubes and figure out how you can get comfortable under the sink, and then follow these instructions to install the riser tubes. Because the stainless braided type of risers don't require cutting to length, we walk you through installing plastic and copper riser tubes.

Special instructions for a pullout faucet

Most pullout faucets have a single handle. In addition to the shanks, you have to attach a nut and large washer to the main faucet body. When the nut is tightened, the faucet is secure. The pullout faucet has two tubes with threaded fittings extending from the body to which you attach the supply tubes.

After you mount the faucet, insert the tube of the pullout wand through the faucet base and attach the hose nut to the faucet. Some nuts are self-sealing. For those that aren't, wrap Teflon tape around the male threads before attaching the nut.

Check the hose by pulling the wand out of the base to make certain that it isn't twisted. You may have to twist the wand to remove any kinks.

Some faucets are unique and may require a special installation technique that we don't have room to cover. For that reason, always follow the manufacturer's instructions.

Trimming the riser tubes to length

The faucet comes with coupling nuts that screw onto the ends of the hot and cold tailpieces. If necessary, remove these nuts from the faucet or find them among the faucet parts.

1. **The top end of the riser has a preformed head. Slip the coupling nut over the other end of the tube, put the riser against the bottom of the tailpiece and thread on the coupling nut — see Figure 15-5.**

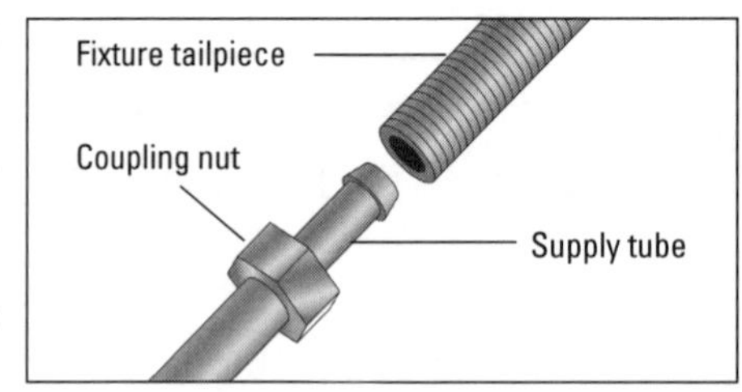

Figure 15-5: The preformed end of the riser tube inserts into the end of the faucet tailpiece.

   Tighten the coupling nut by hand. Repeat with the second riser.

2. **Push the loose end of the riser next to the stop valve and use a pencil to mark its length on the tube.**

   Be sure to mark the tube so that it's long enough to extend fully into the stop valve. Mark both the hot and cold riser tubes.

3. **Remove the tubes and cut them to length.**

   Cut plastic with a sharp knife, copper with a hacksaw or a tubing cutter.

Installing the riser tubes

All types of riser tubes are held in place with compression fittings — a compression ring that you squeeze around the riser tube. Overtightening the fitting can distort the ring, forcing it to cut into the tubing and cause a leak. To avoid crushing the pipe, stripping the threads, and breaking the plastic fittings, thread the nuts on the finger tight. Use slip-joint pliers, a basin wrench, or 6-inch end wrench to gently tighten.

1. **Reinstall the riser tubes to the valve tailpieces and tighten them by hand.**

2. **Slip the compression nut from the stop valve over the end of the riser tube, then slip on the compression ring — see Figure 15-6.**

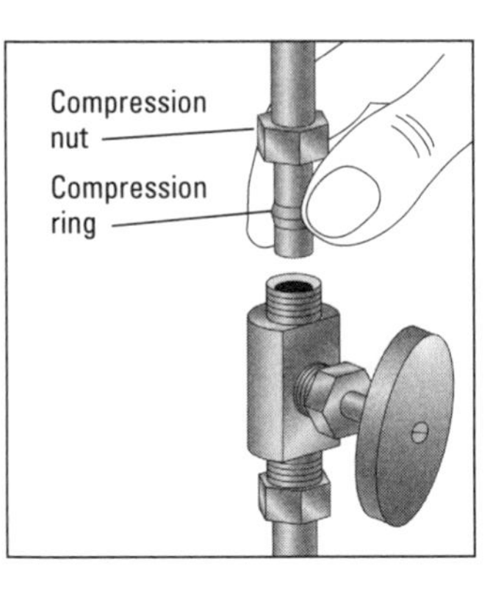

Figure 15-6: Slide the nut and compression ring on the end of the riser tube before you insert it into the stop valve.

3. **Place a dab of caulk on the compression ring and tubing, and push the tube into the base of the stop valve.**

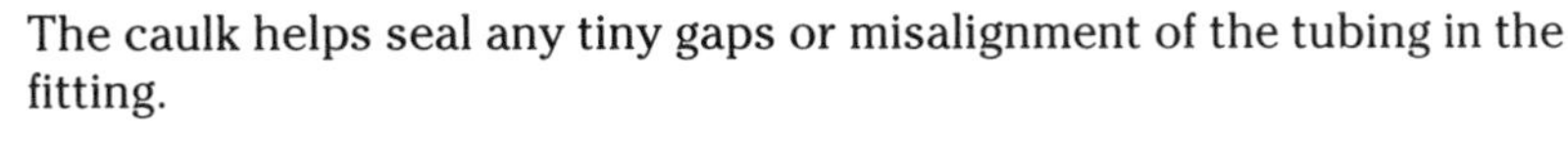

The caulk helps seal any tiny gaps or misalignment of the tubing in the fitting.

4. **Tighten the compression nut by hand.**
5. **Go back and tighten the coupling nuts on the faucet with a basin wrench.**

   You only have to turn these nuts about a ½ turn to tighten.

6. **Tighten the compression nuts on the stop valves.**

   Tighten plastic tubes ¼ to ½ turn. Tighten copper tubes a full turn. Tighten the nuts on stainless steel-mesh tubing about ¼-turn past hand tight (or until they feel snug).

Testing your handiwork

If you removed the p-trap to give yourself more room, now is the time to reinstall it (see Chapter 13). Recheck all of your fittings to ensure that all of the connections are made and tight. Open one shutoff valve and check for leaks. If you don't see any leaks, open the other valve and check again.

If you do see some drips from the compression nuts on the stop valve, give the nut a ¼ turn.

Replacing a Bathroom Sink Faucet

You may be half asleep when you turn on the bathroom faucet every morning, and may not be aware of its full power. But just wait until the day you turn or pull it on and nothing happens — or just as bad, it begins to leak. Then, you notice it big time.

When you need to replace a worn vanity sink faucet with a new one, you'll be glad to know that it's a doable project and one that also dresses up your bathroom.

Removing the old faucet

Begin by removing all of the supplies stored under the sink. If you're planning to replace the sink, leave the faucet in the sink because it's easier to work on after it's out of the cabinet.

The key to successfully installing a new two-handle faucet is understanding the distance between the holes of the existing faucet. After you remove the old one, measure the space between the holes and choose a similar one at your local plumbing or hardware store (see the "Purchasing a new bathroom faucet" section, later in this chapter). This isn't a concern if you have single-handle faucet because most only have one hole — and therefore no distance to measure — with this type of faucet. Other single-handled faucets are designed to mount over multi-hole sinks.

- A *bottom-mount faucet* has only the control handles and spigot mounted on the top of the vanity or sink while the valve body is under the countertop. This type of faucet seldom has a continuous base plate but has the control handles and spigots sitting individuating on the countertop. The valves and spout are usually held in place by threaded packing nuts or *escutcheons* (cover trim pieces).
- A *top-mount faucet* is probably the most popular style in use today and has all of the functioning parts installed above the countertop, usually covered by a base plate. Both the control handles and the control valves are mounted above the sink while the tailpieces protrude through the mounting holes in the countertop.

Bottom-mount faucet

Here's how to remove a bottom-mount faucet:

1. **From under the sink, turn off the water supply to the faucet and remove the riser tubes leading to the faucet (see Chapter 13).**
2. **Use a screwdriver to remove the handle.**

   You may have to remove a decorative cover, called an *escutcheon,* to expose the mounting screw. Loosen the screw and pull off the handle.
3. **Use slip-joint pliers to loosen the nuts or escutcheon from which the valve stem protrudes.**

You may find that the escutcheon is merely decorative and can be pried up to reveal the nut.

4. **As you loosen the packing nuts or installation nuts, you can remove the faucet from the underside of the sink.**

Top-mount faucet

You remove a top-mount faucet this way:

1. **From under the sink, turn off the water supply to the faucet and remove the riser tubes leading to the faucet (see Chapter 13).**
2. **Use a basin wrench to remove the supply tube nuts, as shown in Figure 15-7.**
3. **Remove the fastening nuts from the tailpieces with a basic wrench or slip-joint pliers.**
4. **If the nuts are badly corroded or rusted, use a hammer and chisel or a hacksaw to loosen the nuts that are fused to the bolts.**

   This usually means that you can't reuse the faucet.

Be careful swinging the hammer — a misdirected hammed blow can easily chip or crack the sink.

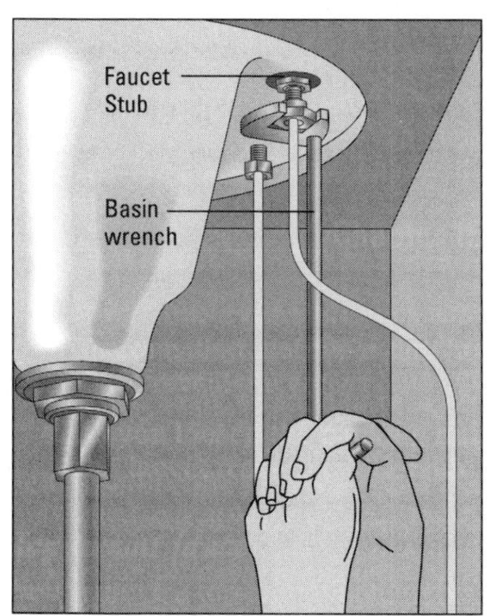

Figure 15-7: Use a basin wrench to loosen the nuts on the supply tubes and remove them. Then, loosen the nuts that hold the faucet in place.

Purchasing a new bathroom faucet

Shopping for bathroom faucets is fun, but you may need to spend a lot of time to find just the right one for your bathroom. Shop at hardware stores, in home centers, at plumbing suppliers, and on manufacturers' Web sites (see Chapter 21). The faucet manufacturers have been working overtime to provide a large selection at a variety of prices. Many "Euro" design faucets have recently made their way into our stores.

Consider your budget before you begin shopping. Faucets with plastic bodies can be purchased for as little as $50. For a quality faucet, expect to pay around $100, while a top-of-the-line faucet can cost more than $300.

In addition to the features list in the "Purchasing a new kitchen faucet" section, earlier in this chapter, here are some of the features you can find:

- **Mounting hole options:** You can find narrow spread (4-inch centers), wide spread (8-inch centers), adjustable (6- to 12-inch centers, some that can be squeezed into 4-inch centers), and single-hole models. Choose the model that appeals to you, is the most convenient, and — most importantly — fits the sink you own. If you're buying a new sink, or installing a new vanity counter, coordinate your sink and faucet selection.

  With faucets that have a 4-inch center, the space between the handles and the spout is limited, especially if the faucet has large, round handles. This tight space makes them more difficult to clean.

- **Handles:** Some single-handle faucets have the handle mounted directly behind the spout. Others have the handle mounted to one side of the spout. You can find many choices of handles. In fact, some makers don't supply handles with the faucet — you select your favorite style for an additional charge.
- **Spout length:** The length of the spout varies a great deal. Short spouts don't take up much room, but they can dribble water down the side of the basin, barely getting the water into the sink bowl. Other spouts are longer and still others are *raised,* in which the spout arches upward so it can reach over large pots. If you're going to mount the faucet in the counter, measure how long a spout you need before purchasing a faucet.
- **Pullout spout:** Pullout spouts with push-button sprayers are great for rinsing your hair, bathing a baby, or for filling a container near the sink.

Installing the faucet

Follow these steps to install a two-hole faucet.

1. **Unpack the new faucet, lay out the parts, and read the installation instructions carefully.**

Unless you've installed an identical faucet recently, you'll probably find some useful information and save yourself from incorrectly installing the faucet.

2. **Set the faucet on the base gasket and through the holes (see Figure 15-8).**

   Not all faucets use gaskets. Instead, your instructions may recommend using plumber's putty or silicone caulk around the base of the faucet. We suggest that you set the faucet without putty or caulk; then seal the base later.

3. **From under the sink, thread the locknuts onto the faucet tailpieces and draw them up finger tight.**

   Align the faucet base so it is parallel with the back of the sink, as shown in Figure 15-9. Tighten the nuts with pliers or a basin wrench. If the faucet has plastic nuts, tighten them a little past finger tight with pliers in order to ensure a tight fit.

   A single-hole faucet uses a single, large nut threaded onto the main faucet body. The mounting principle is the same; there's just one nut.

Many self-rimming sinks don't have a place to mount a faucet, so you install it in the counter top. If you've installed a new counter and are using one of these sinks, you have to drill the mounting holes for the faucet. Make sure you locate the hole close enough to the bowl so that the water flows into the bowl.

Cultured marble tops with integral sinks work essentially the same way. You mount the faucet in the counter, but they usually have holes already drilled for the faucet. If the holes aren't drilled, the same precautions apply.

Adjustable faucet sets (which can be installed in a variety of sink hole configurations) come with components that you assemble on the job. Use them in countertops, in wider center sets in sinks, and in 4-inch centers (although it's sometimes a squeeze). The spout and valves are installed separately, while the connecting tubes are attached to a tee fitting to allow water to flow from the valves to the spout. The connecting hose nuts have rubber seals or are self-sealing.

Purchasing traps and riser tubes

Some faucets can be purchased complete with matching drain, popup assemblies, stop valves, and riser tubes. If you're planning to replace the sink, consider such a purchase. If not, purchase a new set of riser tubes with the faucet (see the "Selecting a trouble-free supply [riser] tube" section, earlier in this chapter).

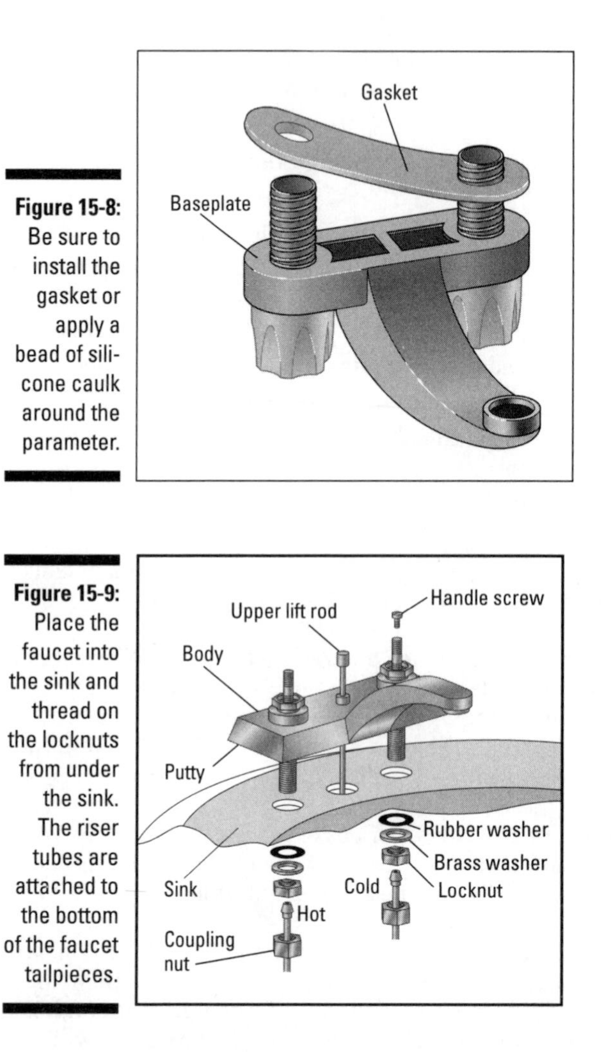

Figure 15-8: Be sure to install the gasket or apply a bead of silicone caulk around the parameter.

Figure 15-9: Place the faucet into the sink and thread on the locknuts from under the sink. The riser tubes are attached to the bottom of the faucet tailpieces.

Installing the pop-up and p-trap

A new pop-up assembly is easy to install. You could leave the old drain assembly in place, but chances are it's pretty beat up and won't match the finish of your new faucet. Consult the manufacturer's installation instructions and identify the parts of the particular pop-up assembly, shown in Figure 15-10. Styles vary, but these are the basic steps:

1. **Thread the locknut, washer, and gasket onto the drain body.**
2. **Apply a bead of putty to the underside of the flange.**

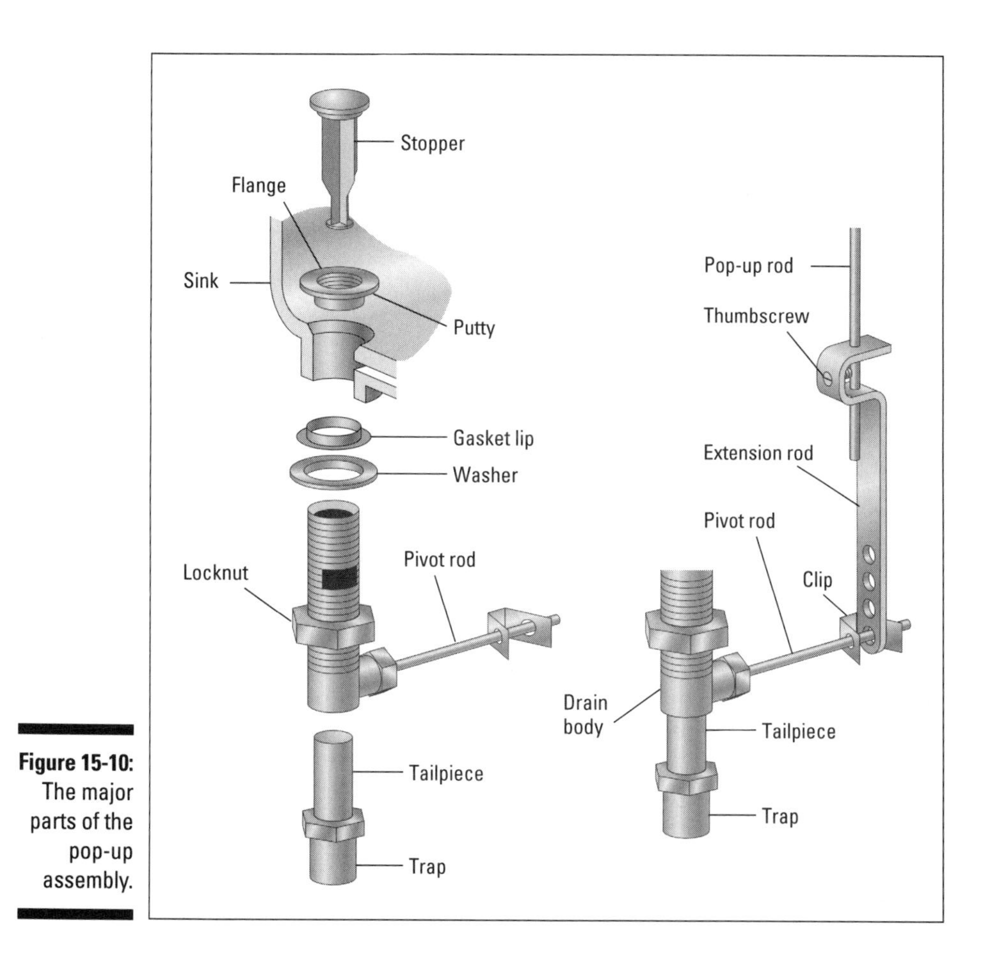

Figure 15-10: The major parts of the pop-up assembly.

3. **From underneath the sink, push the drain body up through the sink drain hole and thread the flange fully onto the body.**
4. **Use slip-joint pliers to tighten the locknut firmly.**

   Hold the drain assembly while you tighten the locknut to keep the pivot rod pointed toward the back of the sink.
5. **Thread the tailpiece into the bottom of the drain body and tighten.**
6. **Position the extension rod onto the pop-up and tighten thumbscrew.**
7. **Drop the stopper into the drain body.**
8. **Reinstall the p-trap.**

   Take the p-trap apart and slip the lower part of the trap onto the tailpiece that protrudes from the sink. Push the upper part of the trap, called the *drain arm,* into the drain in the wall. Reassemble the two parts of the p-trap and tighten all of the slipnuts as shown in Figure 15-11.

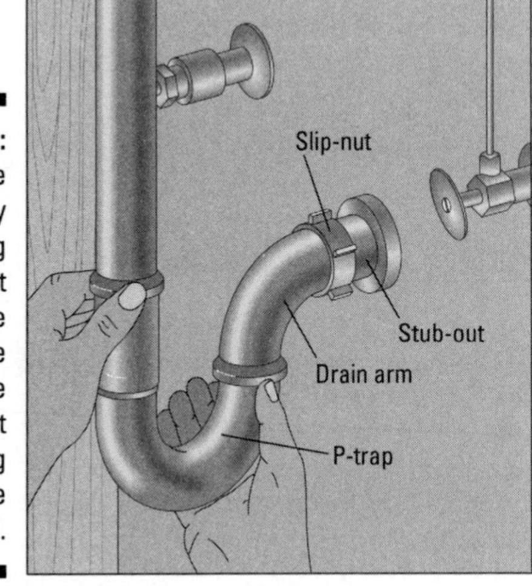

Figure 15-11: Install the p-trap by threading the slipnut onto the tailpiece and the stub-out protruding from the wall.

Hooking up the water

Congratulations, you're almost finished! Connect the faucet tailpieces to the water supply, but be gentle — installing riser tubes requires a light touch. The greatest danger is overtightening a connection. To avoid this, thread the nut finger-tight at both ends. This prevents crushing the pipe, stripping the threads, and breaking the plastic fittings. Use slip-joint pliers, a basin wrench, or a 6-inch end wrench for tightening.

1. **Tighten copper compression nuts about two turns past finger-tight.**
2. **Tighten plastic tubes just past snug.**

   If they're too tight, you can damage and break off the formed end.

   For plastic compression fittings, turn the nut until it feels tight, and then give the pipe a tug to ensure that it can't be pulled out.
3. **Tighten the nuts on steel-mesh tubing about ¼ turn past finger-tight or until they feel snug.**
4. **Apply a bead of silicone tub and tile caulk around the base or escutcheon.**

   The caulk is available in many colors. Select a color that matches the sink or faucet color.

Testing your handiwork

Recheck all of the fittings to ensure that the connections are made and tight. Open one shutoff valve and check for leaks. If you don't see any leaks, open the other valve and check again.

When you're certain that you don't have any leaks, you can begin to enjoy the success of a job well done!

Chapter 16

Replacing a Tub or Shower Spout and Faucet

In This Chapter

- Removing the old spout and faucet
- Choosing a replacement
- Replacing the spout and faucet
- Installing a flexible shower adapter

Tub and shower valves are very reliable and seldom need repair, which is good news, because they are buried inside of a wall. Most types can be repaired (see Chapter 11), but you may decide that it's time to replace really old fixtures. This chapter shows you how.

Before You Begin

The real challenge in replacing a tub or shower valve is getting access to the valve without destroying the wall. So, the best time to replace these valves is during a major bathroom remodeling when walls are being replaced. Otherwise, if in order to replace the valve you have to rebuild and retile the enclosure around the tub or shower, this job can get expensive.

Some house builders have anticipated problems with the plumbing coming up sometime during the life of the houses they build, and make the tub and shower valves accessible through removable panels. Depending on the floor plan of your house, these inspection panels may be located in a hallway or closets. If your house does have a panel like the one shown in Figure 16-1, you can usually replace the valve without disturbing the wall in the bathroom or shower.

Figure 16-1: A removable inspection panel allows access to the plumbing behind the bathroom wall.

Many of the single-handle tub and shower valves have a large *escutcheon* (trim cover) that, when removed, exposes the valve body. This type of valve can be serviced and even replaced by working through the mounting hole.

Another factor to consider before taking on this job is the type of valve you are replacing. Very old valves hooked up to old galvanized pipes are particularly difficult to work with. If the old valve is installed with union fittings (see Chapter 4) between the hot and cold supply pipes, the valve is fairly simple to replace. Union fittings allow the valve to be removed without cutting the pipes.

So if you can get at the valve without having to cut up your wall and the valve is installed with plastic or copper pipes (or you have unions in the galvanized supply pipes), this is a pretty straightforward project.

Removing the Old Valve

If you don't have shut off valves for the shower, you have to turn off the water supply throughout the house — plan accordingly and tell family members that life as they know if will cease to exist. Use heavy cardboard or old dropcloths to protect the surface of the tub while you work on the faucet.

Assuming that you can get to the valve through a service panel, removing the old valve is rather straightforward. Of course you can cut into the bathroom wall if you're really determined to get this sucker out (see Figure 16-2).

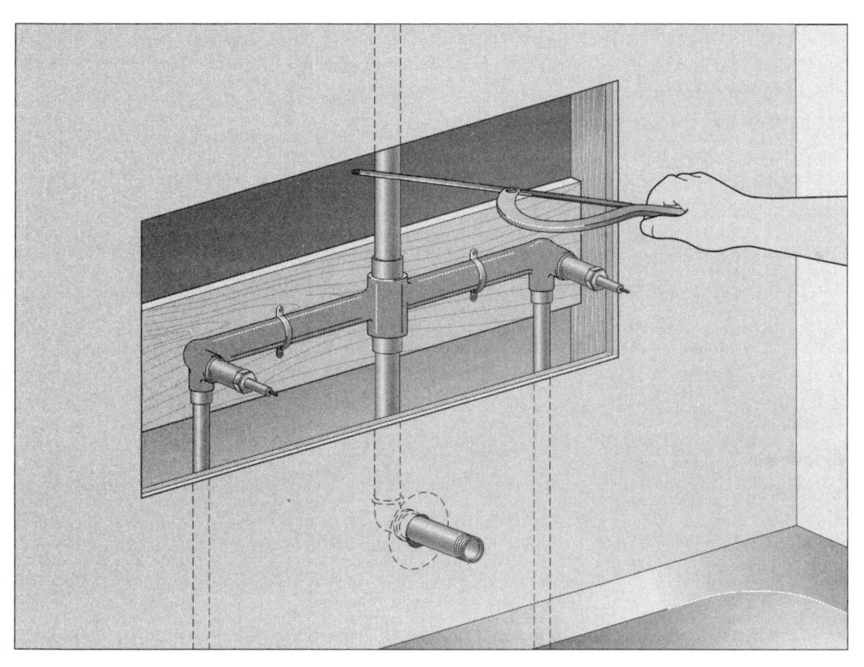

Figure 16-2: If your house doesn't have an inspection panel, consider cutting into the wall to expose the shower valve.

Here's what you do:

1. **Turn the water off.**

   Few of the older tub or shower valves were not installed with shut-off valves so you will have to turn off the water at the main valve. Open the valve to allow the water to drain.

2. **Remove the handles and escutcheons — see Figure 16-3.**

   From the bathtub side of the wall, remove the handles. Most handles are held in place by a screw in the center of the knob. The screw may be under a cover. Remove the screws and then pull off the hot and cold handles. If there is a *diverter handle* for the valve, which sends water from the tub to the shower, remove it.

3. **Remove the spout.**

   Place the handle of a large screwdriver into the opening in the end of the spout and use it as a lever to turn the spout in a counterclockwise direction. After a turn or so you should be able to unscrew the spout by hand.

4. **Remove the showerhead.**

   Use slip-joint pliers to loosen the showerhead. Because you're replacing it, don't worry about damaging the finish. Also, unscrew the showerhead and pipe coming out of the wall that the showerhead was attached to.

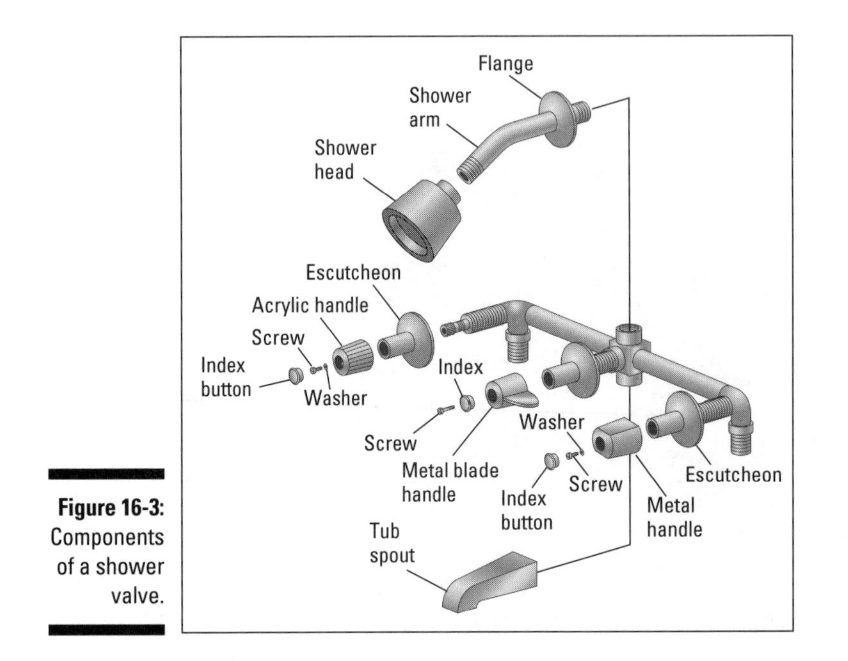

Figure 16-3: Components of a shower valve.

5. **Loosen the union joints.**

   Move to the other side of the wall (see Figure 16-4), and use two pipe wrenches to loosen the unions connecting the valve to the hot and cold water supply pipes. If you don't have unions, you have to cut the pipes with a hacksaw.

6. **If you have copper or plastic pipes or you don't have any pipe unions in galvanized pipe, use a hacksaw to cut these pipes a foot or so below the valve.**

   Cut the hot and cold pipes carefully so you get a nice, square end. A little bit of extra time spent cutting carefully make the pipes easier to solder or to have a new section glued on. (You don't have to be too careful cutting the galvanized pipe because you have to replace that section of pipe.)

7. **Remove the valve body and any mounting hardware.**

   The valve body may be screwed to the wall framing or held in place by pipe straps. Remove the hardware and body.

8. **Save the pipe that leads up to the showerhead.**

   The top of this pipe may be held in place by a fitting called a *drop ear elbow.* You can reuse this pipe.

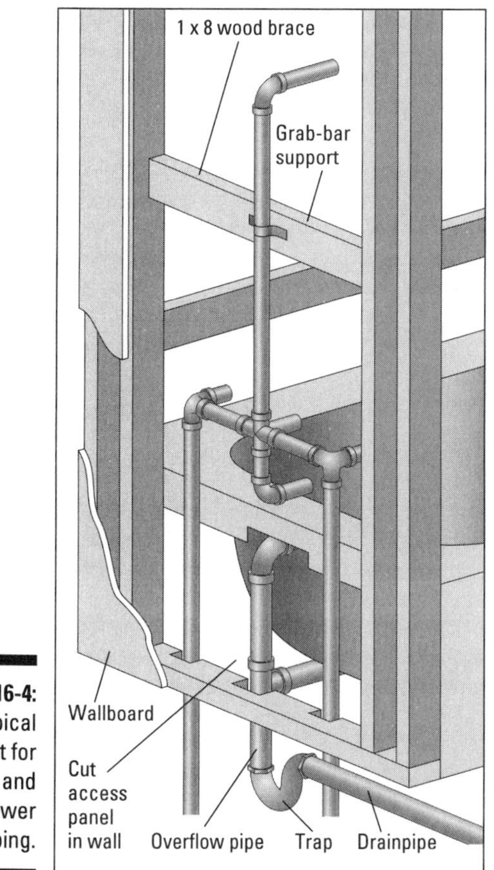

Figure 16-4: A typical layout for bath and shower plumbing.

Choosing a Replacement Valve

Unless you are going to make modifications to the wall, the type of valve you remove dictates the style you purchase. While you're checking out the different valves, take a close look at the installation instructions. Some valves are designed to have the supply pipes soldered directly into the valve. Others require threaded fittings.

You'll find a wide selection of transition fittings in copper, plastic, and galvanized steel. The fitting are designed to allow you to attach different piping systems to the new valve. Transition fittings allow you to connect copper pipe to existing galvanized pipe and the shower valve. If you have a propane torch and a tubing cutter, working with copper pipe isn't difficult (see Chapter 4), The same fittings are available in plastic, so if soldering isn't in your bag of tricks, plastic may be your best choice.

If you're good at measuring, you can have the store cut and thread the necessary galvanized steel pipes to install the faucet, but we don't recommend this approach. Working with galvanized steel pipes is difficult!

Replacing the Valve

Most older valves have a separate hot and cold handle; some have a third handle between the hot and cold to direct the water flow to the tub or shower head. These valves can usually be repaired — see Chapter 11. If your valve can't be fixed or you want to modernize your bathroom, follow the directions in this section to remove the old valve and install a new one. Like most plumbing fixtures, the valve comes with specific installation instructions — be sure to follow them.

Basically, you install the new valve by reversing the steps you took to remove the old one, using Figure 16-5 as your guide.

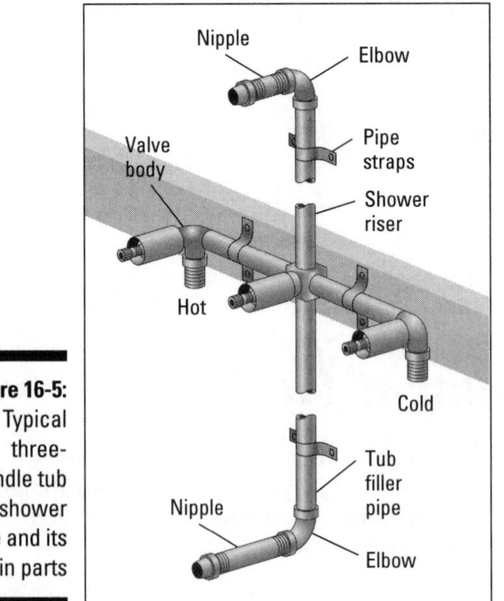

Figure 16-5: Typical three-handle tub shower valve and its main parts

1. **Install the valve.**

   Put the valve in position. You may have to enlarge the opening in the wall to accommodate the new valve stem body because it may be larger in diameter than the one you replaced. Mounting hardware is supplied by the manufacturer.

2. **Install the transition fittings.**

   Install the transition fittings on the ends of the old piping. If you're working with galvanized pipe, use a fitting that screws onto the end of the pipe and provides the proper connector for either plastic or copper on the other end. If you have copper or plastic piping, a simple coupling fitting joins the old pipe to the new.

3. **Run the hot and cold piping to the valve.**

   Starting at the transition fitting, cut and fit the pipe to the valve. Cut the pieces of pipe for both the hot and cold. If you're using copper, disassemble the pipes and carefully clean the ends of the pipes and inside the fittings. If you're using plastic, mark the fittings and pipes with a marker so that you can reassemble the pipes quickly as you glue the parts together.

4. **Install the tub spout.**

   You may be able to use the old pipes that lead from the valve to the tub spout. The length of these pipes is given in your installation instructions. If you can reuse these pipes, apply pipe dope to the threads and install the pipe with a pipe wrench. Otherwise cut and install new plastic or copper piping (see Chapter 4).

5. **Install the shower riser pipe.**

   The shower *riser* (supply tube) is attached to the top of the valve. The challenge you face here is that you can't turn the pipe if it's attached to the wall with a drop ear elbow (see the "Removing the Old Valve" section, earlier in this chapter). This isn't a problem with plastic or copper, but can be a struggle with galvanized steel.

   Look through the hole in the shower wall where the shower arm pipe came out. If you see an elbow that's nailed or screwed to the wall framing, you can't pull this pipe out of the wall cavity. Try to loosen the screws or nails with a long screwdriver, which allows you to rotate the shower riser pipe and thread it into the new valve.

6. **Install the escutcheons and handles — refer to Figure 16-3.**

   The valve comes with instruction on how to install the trim rings that cover the escutcheons. Follow these instructions to complete the installation of these parts and the showerhead.

7. **Test valve for leaks.**

   Open the shower valve and then turn the water back on. As soon as the water flows out of the tub spout, turn off the water. Check the joints between the valve and new piping. Turn on the water and start the shower. If the diverter is in the tub spout, check for a leak in the pipes leading to the spout and the shower head.

Installing a Flexible Shower Adapter

If only everything in life were this simple: For about $25, you can buy a flexible shower adapter kit and transform a showerless bathroom into one with a shower. The kit, shown in Figure 16-6, includes the flexible shower hose and a tub spout with a hose outlet on the bottom of the spout. You replace an existing spout with the new one, attach the flexible shower hose and head and voilà – instant shower. Some kits don't adapt to older spouts, so this trick may not work for everyone, but it's a quick fix that's worth investigating.

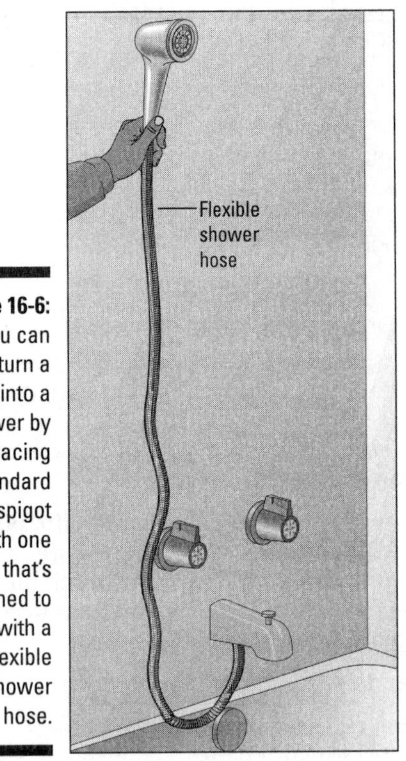

Figure 16-6: You can easily turn a tub into a shower by replacing the standard tub spigot with one that's designed to work with a flexible shower hose.

You'll need to have the following tools and materials:

- Adjustable wrench or locking-type pliers (see Chapter 3)
- Two pipe wrenches (see Chapter 3)
- Electric drill with glass and tile bit (if drilling into tile)

- Screwdriver
- Masking tape
- Masonry anchors (if drilling into tile)

Read the directions included with the shower adapter kit to become familiar with the parts and installation steps. Follow these steps to install a flexible shower adapter:

1. **Turn off the water at the main supply line.**
2. **Remove the old tub spout by twisting or unscrewing it.**

   Wrap a rag around the spout to protect it while removing it. (See the "Removing the Old Valve" section, earlier in this chapter, for more information on removing a tub spout.)
3. **Find the adapter hose outlet on the bottom of the new spout. Thread the hose coupling onto this threaded stub, and then use an adjustable wrench or locking-type pliers to tighten it.**
4. **Stand in the tub and locate the showerhead hanger so that is casts a fine spray at the correct height.**

   Hold the flexible hose up to the wall and mark the best location for the holder. The hose is stored in the holder and should be able to be lifted freely from it.
5. **Mark the location for the holder with a washable marker and then insert wall anchors into holes.**

   Secure the holder with the screws provided.

   To attach the holder in a ceramic tile wall, use an electric drill with glass and tile bit, masking tape, masonry anchors.
6. **Insert the shower hose in the holder.**

Chapter 17

Replacing a Toilet Seat and Toilet

In This Chapter

- Replacing a toilet seat
- Removing the old toilet
- Selecting and installing a new toilet

Your toilet may just need a new seat — or maybe the whole dang thing needs to be replaced. Whatever your toilet project, this chapter helps you do the job. Now, we admit that a lot of people are intimidated by the thought of removing a toilet — yikes, what will happen? Perhaps you can hardly imagine your bathroom without it. But put yourself in our capable hands — we take you through all of the steps of removing and replacing first the ol' seat, and then the big kahuna itself.

Replacing A Toilet Seat

Replace a toilet seat and cover for any number of reasons: it wobbles, it's ugly, or you just want a new one. Believe it or not, replacing one is easy to do.

You may not get much help when shopping for a replacement seat and cover (they come as a unit), because the home centers and bath departments of discount stores that we visit offer virtually no assistance in finding the appropriate size seat. The key is knowing the distance between the bolts — usually about five to six inches — and the length of the seat itself. We have devised two approaches to finding the right seat for your throne.

- Remove the old seat, wrap it in a sack or bag, and take it to the store to find one with the same dimensions. When you get to the store, remove your old seat from the bag and put it on top of the various seats that are on display until you find one that matches.
- Find the name of the manufacturer on your toilet and go to their Web site, where you can order a replacement or find a local retailer that stocks one.

Removing the old seat

Removing the old seat can be a five-minute job or it can turn into a marathon project, depending on the state of the mounting bolts that hold the seat to the toilet.

Cooperative seating

Assuming the bolts operate easily, removing the seat is easy. Here's how:

1. **Close the lid on the seat and find where the seat and cover attach to the toilet bowl.**

   The seat and cover are usually attached with bolts that are concealed by two hinged caps or *tabs*.

2. **Open the tabs by prying them up with the blade of a screwdriver to expose the seat bolts.**

   You probably have to reach under the back of the toilet and hold the nuts to keep them from turning. Some seats have the bolts molded into the *hinge bracket,* which is attached to the toilet seat, so that all you have to do is loosen the bolts.

3. **Lift off the old seat and clean the area beneath the hinges. Give the toilet bowl a thorough cleaning with a toilet cleaner and brush, and use a household cleaner to wash the outside surface of the toilet.**

A #!%* stubborn seat

If the bolts are corroded or difficult to loosen, a little bit of penetrating oil or a hacksaw will eventually allow you to conquer those bolts.

- Soak the bolt in a penetrating like WD-40. Let the oil soak in for a few hours or even overnight, and then try loosing the nuts with a slip-joint pliers (see Chapter 3). To keep the heads of the bolts from turning, insert a large screwdriver into the slots in the bolts heads.
- If that's a no-go, use a hacksaw to cut through the bolts. Almost all seats use brass bolts, which aren't too difficult to cut with a hacksaw — see Figure 17-1. If the hinge bracket is plastic, you may find it easier to cut through the bracket and bolt from the top of the toilet. If the hinge is metal, you have to cut the bolts loose from underneath.

Installing a new seat

To install a new seat, read the instructions to see if any tools are needed and to understand what's involved. Most toilet seats have brass or plastic bolts and plastic nuts so they're easy to install. Basically, this is how to do it:

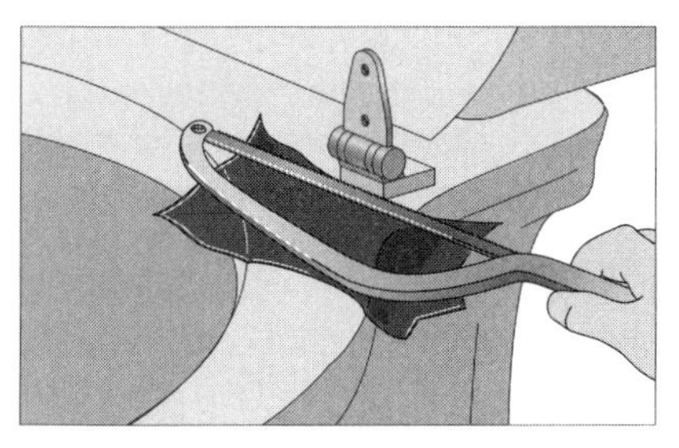

Figure 17-1: Corded brass or rusty steel bolts must be cut through with a hacksaw.

1. **Open the tab covers and position the new seat on the toilet bowl by lining up the tab holes for the bolts in the back.**
2. **Insert the bolts in the holes and fasten them, either with your hands or by using a wrench.**

   Don't over tighten the bolts — a half turn with a wrench is usually all that you need to secure each one.

Does your toilet seat slip to one side? If so, tighten the toilet seat bolts to secure it. Use a wrench or pliers to adjust the bolt on the side that's slipping. Open the tab cover to access the bolt and then tighten it down.

Choosing a New Toilet

Today's toilets are better looking and more efficient than ones of years past, because they use a mere 1.6 gallons of water per flush, compared to 3 to 5 gallons. You can find two basic styles of toilets: a one-piece unit with a combined tank and bowl, and a two-piece unit with a separate water tank and bowl. Two-piece units are more popular and less expensive than one-piece toilets.

The most popular flush mechanism is a gravity-fed flush unit that uses the force of gravity and a siphoning, pull-through action to empty the bowl. Another type, a pressure-assisted toilet, relies on either compressed air or water pumps to boost the flushing power. Our choice for a budding do-it-yourself plumber is the two-piece gravity-assisted toilet.

Most toilets are designed with a standard 12-inch *rough-in dimension,* which means that it's located 12 inches from the wall — see Figure 17-2. You can also find toilets with 10-inch or 14-inch rough-in dimensions.

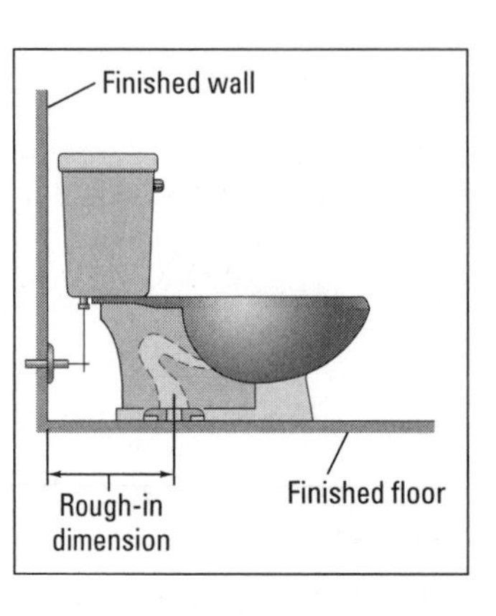

Figure 17-2: Measure the rough-in dimension: the distance from the closet bolts to the finished wall. Purchase a new toilet with the same dimension.

If you're replacing a toilet with a wall-hung tank, it will have a rough-in dimension of 14 inches. If this is a case, you can install a toilet designed for a 12-inch rough-in, but it'll end up a couple inches from the back wall. Place a short piece of 2x4 lumber on the wall behind the tank to give support to the tank.

If your old toilet has a 10-inch rough-in dimension, you can purchase a new toilet with this dimension — you may have to special order it, however.

Replacing Your Toilet

Before you begin this project, make arrangements for using another toilet. If your household has only one toilet, this may be the most important part of the job! Find a friendly neighbor (who is indebted to you) and prevail on his hospitality. Of course, if you have another toilet in the house, you're in good shape.

You'll probably find the bathroom easier to work in if you remove any extra doodads or decorations. Remove towels, baskets — whatever's in the room — so that the toilet is free and clear to work on.

Removing the old toilet

Follow these steps to remove the old toilet:

1. **Pour ¼ cup of bleach or toilet bowel cleaner into the toilet and then flush the toilet a few times.**

Should you replace a toilet yourself?

You may be worried about what you'll find if you remove the old toilet or that water will shoot out if you break the toilet as you remove it. If you empty the tank and bowl of water, you won't have to worry about a water spout occurring in your bathroom.

The toilet is held in place by a couple of bolts that may resist removal because they've been in place for a long time. Other than this challenge, replacing a toilet may seem a lot more mysterious than it is difficult.

2. **Turn off the water to the toilet and disconnect the water supply line to it at the bottom of the toilet tank. Flush the toilet again.**
3. **Carefully lift the top of the tank and set it aside.**
4. **Use a large sponge to remove the remaining water in both the toilet tank and bowl. Soak up the water in the tank with the sponge and squeeze it into a bucket. Use a cup to bail the water out of the bowl and then sponge up the remaining water.**

   The toilet bowl can contain harmful bacteria, so when you're removing the water from the bowl, wear rubber gloves. Wash your hands thoroughly afterward.

5. **Remove the tank from the toilet by loosening the *tank mounting bolts and nuts,* located on the underside of the toilet base, where the tank rests.**

   You may need a wrench to hold the head of these bolts, in order to keep them from turning while you loosen the nuts.

6. **Remove the bolts and nuts (called *closet bolts and nuts*) that hold the toilet to the floor — see Figure 17-3.**

   You can find one bolt on either side of the base of the toilet, and they're usually covered with a plastic cap or cover that lifts off to expose the closet nut. Use a wrench to loosen the nuts and unscrew them.

   Keep some rags on hand to catch any water. Spread newspapers on which to place the toilet.

7. **Remove the toilet.**

   Try to keep the toilet level as you lift it off the floor, then carefully place it on some newspapers. (The bottom of the toilet is dirty, so you don't want to place it on the floor.)

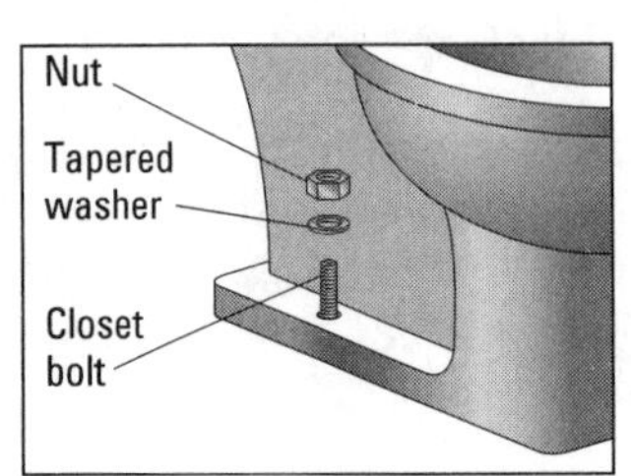

Figure 17-3: Pry off the covers and remove the nuts from the closet bolts to free the toilet from the floor.

8. **Stuff a rag into the hole in the floor.**

   This hole, called the *flange hole,* is a direct path to the soil pipe leading to the sewer or septic system. Usually, the water in the toilet bowl prevents sewer gases from entering your home, but with the toilet gone, you must plug up this opening.

9. **Clean the floor area under the old toilet and use a wide blade scraper to remove the old wax ring and its residue from the floor.**

Installing a new toilet

The job involves anchoring a new toilet bowl to the floor with bolts, and mounting the water tank onto the bowl. Here's what's involved to install a two-piece toilet with a tank and bowl. If you plan to install a one-piece toilet, you can skip the tank installation steps.

To complete this job, you need the following:

- The new toilet and all of its mounting bolts and nuts
- Plumber's putty and a putty knife (see Chapter 3)
- A wax ring and sleeve (to seal the joint between the toilet and the soil pipe leading to the sewer system)
- Adjustable wrench (see Chapter 3)
- Screwdriver and carpenter's level (see Chapter 3)
- Latex tub and tile caulk

Prepping the floor

You must first clean off the *closet flange* (cast iron, copper, or plastic material with slots that hold the closet bolts), because it probably has old wax or plumber's putty on it. Then, attach the new closet bolts to the flange by turning the T-shaped head of each bolt so that it aligns with the flange slot and push the bolt below the flange. Slide the bolt into position and then turn so the head so that it can't be pulled out of the flange — see Figure 17-4.

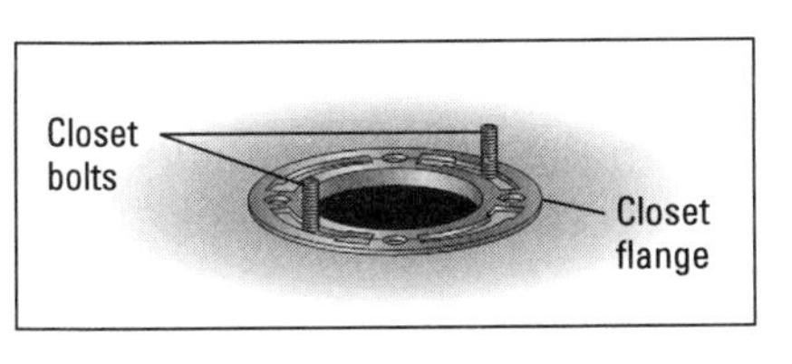

Figure 17-4: Install new closet bolts onto the closet flange. Align them to be parallel to the wall.

Install the other closet bolt in the same way, and then adjust the position of both bolts so that they're the same distance from the wall and are aligned parallel to the wall.

To hold the bolts upright and make positioning the toilet bowl on the flange easier, pack some plumber's putty around them.

Installing the base

The joint between the toilet and the soil pipe leading to the sewer system is sealed with a wax ring. These rings are available wherever you purchase the toilet. You will also need a small container of plumber's putty to seal the bottom edge of the toilet.

Be careful handling the toilet; it's strong but can chip and crack if dropped or struck with a hard object (like a wrench).

Here are the basic steps in installing the toilet base:

1. **Lay the toilet base upside down on a soft surface (such as a scrap of carpeting) so you can work on the bottom of the unit.**
2. **In the center of the toilet base is a short spout called the *toilet horn*. Place the wax ring and sleeve onto the toilet horn and press it down so that it sticks in place. See Figure 17-5.**

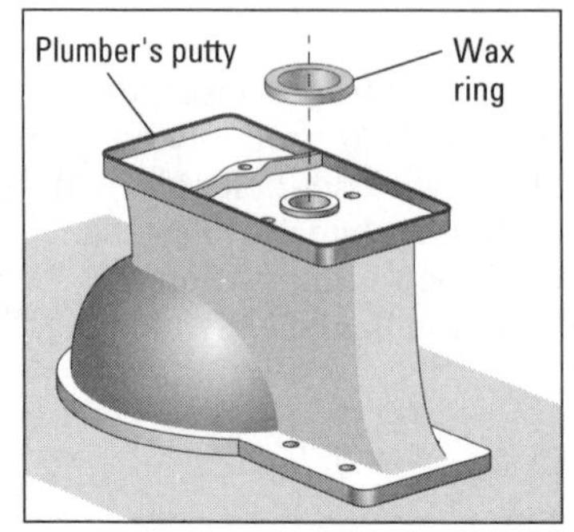

Figure 17-5: Place the wax ring and sleeve onto the toilet horn.

3. **Apply plumber's putty around the bottom edge of the toilet base.**

   Take a handful of putty and work it between the palms of your hands to form a ¼-inch diameter putty rope. Place the rope around the parameter and push it in place so that it sticks to the toilet base.

4. **Remove the rag in the flange hole (hole in the floor).**

5. **Set the toilet bowl on the flange hole by lining up the closet bolts over the holes in the toilet base.**

   It's helpful to have a helper on hand to help you locate the bolts while you're lowering the base onto them.

6. **Gently, but firmly, press the toilet base down on the wax ring.**

   After it is in place, push down firmly to seat the wax ring and force the excess plumber's putty from under the toilet.

7. **Place the washers and nuts on the closet bolts, then use an adjustable wrench to tighten the nuts. Tighten the nuts evenly and check if the toilet is level by placing a carpenter's level across the toilet.**

   Don't over tighten these bolts or you may crack the toilet base.

8. **Use a putty knife to clean away any putty that squeezed out between the bowl and the floor.**

9. **Cover the bolts with the trim caps or tabs that come packaged with the toilet.**

   You may have to cut the bolts off with a hacksaw if they're too long, to allow the caps to cover them.

To hold the caps in place, put some of the excess plumber's putty that you trimmed from the base into the caps.

Installing the tank

Most tanks come with the fill and flush mechanism already installed. The *fill and flush mechanism* regulates the water flowing into the tank — when you push the flush lever, it opens a valve so that clean water in the tank flows into the toilet bowl, flushing out its contents. If this isn't the case with the toilet you have, follow the instruction included with the toilet to install them. (See the "Adjusting the toilet" section, later in this chapter, for information on adjusting the fill and flush mechanism.) When the fill and flush mechanism is installed, follow these steps to install the tank on the toilet bowl.

1. **Turn the water tank upside down and attach the rubber seal (called the *spud washer*) to the pipe that protrudes from the bottom of the tank.**
2. **Carefully turn the tank right side up and center the spud washer over the water intake opening, which you can find near the back edge of the toilet bowl.**
3. **Attach the tank to the bowl.**

   Align the tank bolts and rubber washers with the holes in the tank and insert the tank mounting bolts through the holes in the toilet bowl. Thread on the washers and nuts from the underside of the toilet bowl. (See Figure 17-6.) Hand tighten and then carefully tighten with a wrench until the tank is tight on the bowl — don't overtighten these nuts, however!

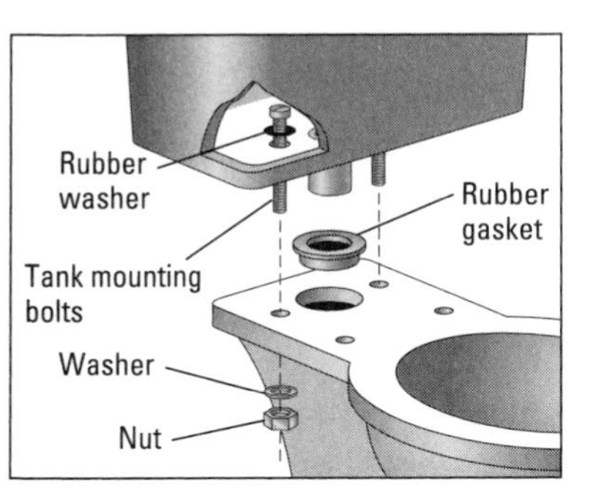

Figure 17-6: Lower the tank onto the toilet bowl, and then install the washers and nuts from the underside of the bowl.

   Some tanks come with the tank mounting bolts preinstalled; others require that you install the bolts. Follow the directions that come with the toilet.

4. **Install the riser tube between the stop valve on the wall and the fill valve protruding from the bottom of the tank. (See Chapter 15 for the lowdown on this procedure.)**

Adjusting the toilet

Fill and flush mechanisms vary according to the manufacturer, so you must follow the specific instructions included with the toilet.

To fill the toilet, turn on the water at the stop valve (see Chapter 15). Check for leaks at the stop valve and fill valve. Carefully watch as the toilet begins to fill. The water flow should begin to slow and stop at the fill line marked inside the tank. If this doesn't happen, follow the manufacturer's direction on adjusting the water level. Some valves have a float with an arm that can be bent for adjustment; others have adjusting set screws.

After the tank fills, flush the toilet a couple of times. Note the action of the flapper valve that's opened by the flush lever. This valve should open and stay open until most of the water drains out of the tank, and then closes. You may have to adjust the chain leading from the flush level to the flapper valve to get this valve to stay open and close completely when the tank is drained of water. If the tank doesn't completely fill, you may have to adjust the position of the tankball rod to get this action correct (see Chapter 12).

Chapter 18

Replacing an Electric Hot Water Heater

In This Chapter

- Sizing the heater to your needs
- Choosing a new water heater
- Removing a water heater
- Installing a water heater with step-by-step instructions
- Maintaining a water heater

You usually have two reasons to replace a hot water heater:

- The interior of the tank has rusted out and is leaking.
- You have a teenager in the house.

No kidding. In an unscientific survey of managers of hot water departments at home centers, the term "teenager" came up time and again when questioned about hot water heaters. It makes sense: A teenager requires thousands of gallons of hot water every day to exist — if the old unit stops cranking out hot water, it's got to go!

People usually never think about a hot water heater until it stops working or starts leaking. It's usually tucked away in a utility closet or down in the basement and is hardly a concern; that is, until you turn on the shower, run the dishwasher or clothes washer, and find that you don't have any hot water. The chapters tells you what to do when that happens.

Of the almost nine million hot water heaters sold each year, about 55 percent are gas units and 45 are electric. If your unit is a gas hot water heater, we suggest that you hire a plumber to replace it. The installation requires electrical skills, plumbing skills, and a licensed plumber to make the final fuel connections to meet local building code requirements. Even though a do-it-yourselfer can make the installation, you must have your gas company or a licensed plumber hook it up to the gas source. Check your local building code requirements for the specific regulation in your area.

But replacing an electric hot water is a doable project for a handy homeowner. The real challenge is moving the old behemoth to make way for the new one. It isn't always easy to maneuver a rusty old tank out of a tight space behind the furnace in the basement and up a narrow staircase without damaging the floor or spraining your back. Another problem: getting rid of the thing. You may have to pay extra to have your garbage service dispose of the unit.

Choosing a New Water Heater

Before removing the old unit, do some investigative work. Look at how the pipes and wiring are connected, and check to see where they lead.

Choose a new unit with the same voltage as the existing unit. Also, select one that has a polyurethane foam insulation — it costs a little more than a standard unit, but because of its improve efficiency, you recoup your investment in lower hot water bills throughout the life of the unit.

Size

The most popular size unit has a 40-gallon tank that comes with a five-year warranty, but you can find a range of sizes, from 30-and 65-gallons. Typically, a 50-65 gallon heater is large enough for a family of four (without a teenager) who don't have a high demand for water.

Table 18-1 gives you guidelines for selecting the right size electric water heater for your house. Refer to this table even if you're replacing an existing hot water heater and you already know the size. You may indeed need a larger size if you run out of water frequently, expand your house to include another bathroom, add new hot water-using appliances like a hot tub or dishwasher, and if your family grows.

Table 18-1: Finding The Proper Size For Your Hot Water Heater

People in the house	*Type of demand*	*Size needed*
5 or more	Regular demand	80 gallon
	High demand	120 gallon

People in the house	*Type of demand*	*Size needed*
3 to 4	Regular demand	65 gallon
	High demand	80 gallon
2 or fewer	Regular demand	40 gallon
	High demand	50 gallon

Features

You can just pick up the phone and order a hot water heater the way you order a pizza, right? Wrong. You may want to check out the features that are available for this hard-working water appliance before selecting one.

The most expensive units last longer, heat more water, and are less costly to operate, and this is reflected in the length of their warranties. When shopping for a hot water heater, look for the following features:

- *Heat traps,* which are installed between the hot water pipes and the water heater to slow the heat transfer from the heater to the pipes. Extra insulation around the unit also helps to prevent heat loss, making the heater less costly to operate.
- *Fast recovery* allows the heater to produce hot water so that it can keep up with peak demand situations: several showers are in use and the washer and dishwasher are operating.
- Stainless steel elements that resist lime build up and accidental dry-fire burnout, in case you accidentally start the heater without any water in it.
- Self-cleaning system to fight lime and sediment build-up, which improves tank life and maintains peak efficiency.

Enlightening reading

Be sure to read the Energy Guide label that spells out typical yearly operating costs (based on national averages) for each model. The label also describes the unit's tank capacity and its insulation r-value. You may also find installation clearances that describe where the unit can be installed (or example, in a closet or alcove), the type of flooring that you can install the heater on, and the minimum distance the you should locate the heater from combustible material like stacks of newspaper or piles of clothes.

Installation kits

While shopping for a water heater, pick up a *hot water heater installation kit,* which contains two flexible sections of copper tubing and the necessary pipe nipples and adapters to complete the installation. This saves you from having to make repeated trips to and from the plumbing department.

Removing a Water Heater

You need the following tools and equipment to remove your old hot water heater and install a new one:

- *Neon circuit tester,* also called a *test lamp,* which is a simple device for determining if an electrical circuit is "live." For under $2, it's a tool that you should definitely purchase.
- Pipe wrenches and adjustable wrench (see Chapter 3)
- Hacksaw or tube cutter (see Chapter 3)
- Screwdriver and carpenter's level (see Chapter 3)
- Bucket
- Pressure-relief valve — you may have to purchase this part if the new heater doesn't come with one.
- Two heat-saver (plastic-lined) nipples (see Chapter 4)
- Wire nuts, which are plastic cabs with threaded insides for connecting the ends of electrical wires.
- Teflon plumber's tape (see Chapter 3)
- Propane torch, solder, and flux (see Chapter 3)
- ¾-inch copper pipe (see Chapter 3)
- Pipe and fittings (see Chapter 3)
- Pipe joint compound (see Chapter 3)
- Flexible water connectors
- *Appliance dolly,* which is a two-wheeled push carriage used to move appliances and other heavy objects. The water heater isn't too heavy when it's empty, but it's awkward to move.

Pay attention to how you disconnect your old water heater — when you install the new unit, you'll connect it in a similar way. To disconnect the old unit, follow these steps.

1. **Turn off the electrical power at the circuit breaker so that no live power goes to the unit.**

   Combining water and electricity is a dangerous duo. Always turn off the power at the main circuit breaker when you're working on an electric water heater. It's not enough to turn off the switch on the water heater.

2. **Turn off the water supply to the heater (see Chapter 2).**

3. **Drain the water heater and disconnect the water pipes.**

   Open the *hose bib* (a drain on the bottom side of the tank), and drain the water into a bucket. If the floor drain is nearby, you can also connect a garden hose to the tank and drain the water directly into the floor drain.

4. **Disconnect the hot and cold water pipes above the unit.**

   Use a pair of wrenches working against each other to disconnect the threaded fitting, as shown in Figure 18-1. If you can't budge threaded pipes, try using a propane torch to heat them so they'll be easier to separate.

   If the pipes are copper (and therefore soldered), you may have to use a hacksaw or tubing cutter to cut the pipes. Make straight cuts through the water pipes just below the shutoff valves.

Figure 18-1: Use two pipe wrenches to open the union fittings to disconnect the water supply pipes from the heater.

5. **Locate the electrical box.**

   The electrical box is located under a cover plate, on the side or top of the heater. Remove the cover of the electrical box to expose the heater wiring.

6. **Disconnect the electrical wires leading to the water heater.**

   Disconnect all of the wires by loosening and then removing the wire nuts that are twisted in the ends of the wire splices. The incoming wire is held in place by a clamp in the cover. Loosen the screws on the clamp to free the incoming wire. Pull the wire out of the box.

7. **Muscle the unit out of its place and load it onto an appliance dolly to remove it.**
8. **Clean up the floor and surrounding area.**

   If a new paint job is needed, do it while the old heater is removed so that you have more space to work with.

Installing a Water Heater

So, you have the bruiser in place and you're ready to install it. This section shows you how to install a new water heater, using a readily available installation kit. But be sure to read the installation instructions that come with the unit before you begin.

1. **Connect the pressure-relief valve.**

   Prepare the new valve by wrapping the threads with Teflon plumber's tape or applying a light coating of pipe dope. Screw the valve into the tank by hand until it's fully seated. Tighten the valve with a pipe wrench. Remember that as you tighten the valve, you want its final position to be so that the overflow outlet faces down. See Figure 18-2.

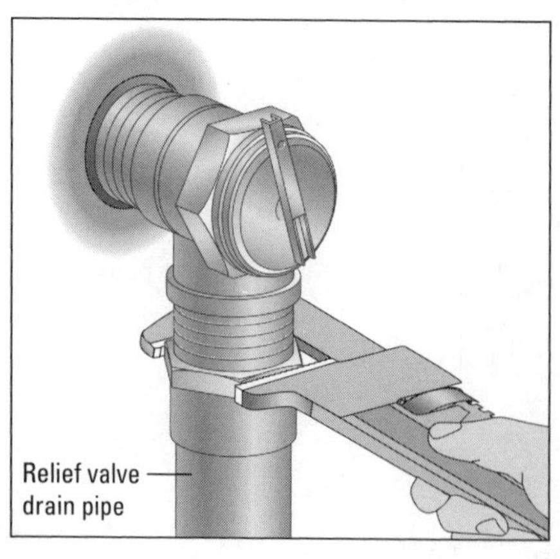

Figure 18-2: Thread the pressure-relief valve into the heater body then install the relief valve drain pipe.

2. **Connect the relief valve drain pipe to the pressure-relief valve using the threaded male adapter.**

   The pipe should extend to within a few inches of the floor.
3. **Wrap Teflon plumber's tape around the nipples that are provided in the installation kit. (You can also use pipe dope.)**

Thread these into the top of the hot water heater. One will be marked for the hot, the other goes in the cold. Use a pipe wrench to tighten them. See Figure 18-3.

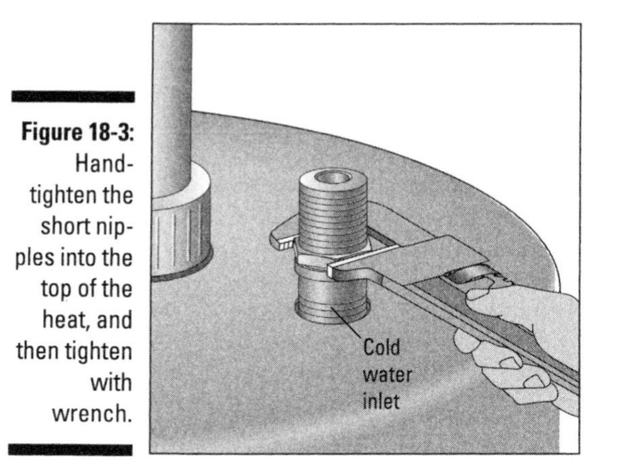

Figure 18-3: Hand-tighten the short nipples into the top of the heat, and then tighten with wrench.

4. **Connect the water lines to the nipples, using flexible water connectors and tightening them with an adjustable wrench — see Figure 18-4.**

   If you have copper or plastic piping, you can use the flexible connectors. To do this you will have to solder (in the case of copper) or glue (in the case of plastic) male fittings on the end of the pipe leading to and from the house plumbing (see Chapter 4). The flexible connectors then thread onto these fittings.

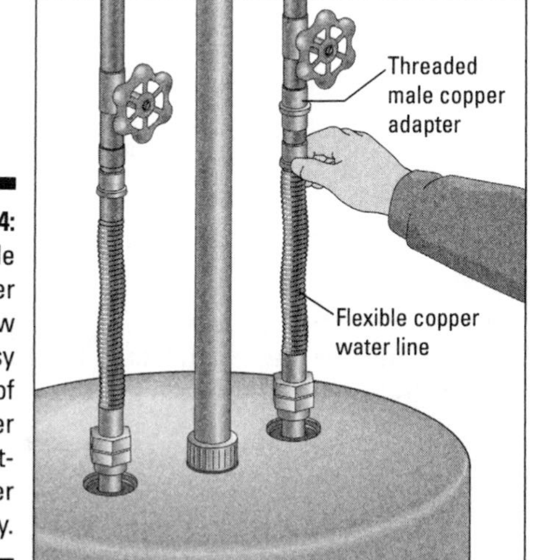

Figure 18-4: Flexible copper tubes allow for easy hookup of the heater to the existing water supply.

Follow the manufacturer's wiring instructions that came with the new water heater. The steps generally look something like the following steps:

1. **Remove the electrical box cover plate. Thread the same wires that you removed from the old heater through the wire clamp attached to the cover plate.**

   If there is only a round opening in the cover plate, you need to purchase a cable clamp that will fit into this ¾-inch diameter hole in the plate. All hardware stores carry this item.

2. **Use wire nuts to connect the circuit wires to the heater wires — see Figure 18-5.**

   Connect the like-colored wires together.

3. **Attach the bare copper or green ground wire to the ground screw and replace the cover plate.**

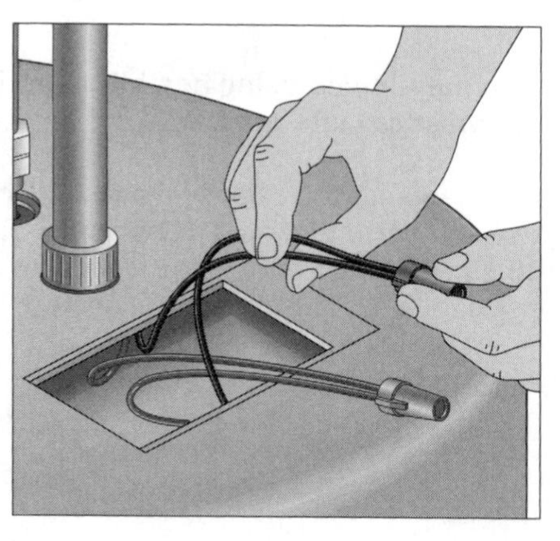

Figure 18-5: Use the supplied wire nuts to connect the black and white wires with the same colored wires from the heater. Ground the heater with the green wire.

4. **Press the reset button on the thermostat and close up the electrical box.**

5. **Open the inlet and outlet valves located in the supply pipes leading to the water heater, as well as all the hot water faucets in the house. Close the faucets when water flows without bursts of air.**

6. **Restore the electrical power to the water heater at the main circuit breaker.**

7. **Adjust the water temperature.**

   Allow the heater to cycle several times, running the hot water until the heater turns on to reheat the water. If after several cycles, the water isn't at the right temperature adjust the thermostat.

   Open the *access panel* on the side of the heater body. (Large heaters have two access panels because they have two heating elements and two thermostats.) Use a screwdriver to set the thermostat to the desired temperature. The temperature settings are marked on the body of the thermostat.

   To avoid scalding injuries, home safety experts suggest 120°. This setting helps reduce damage to the heater from overheating and saves energy.

Maintaining Your Water Heater

To keep a water heater in good working condition, follow these simple steps — once a year for sure, two times is better. The idea is to flush out mineral deposits and sediment that reduce the heating efficiency and cause corrosion.

To flush and drain a water heater, do the following — you need a garden hose and a rag:

1. **Find the cold water supply pipe line above the heater and turn off the valve.**
2. **Hook up the garden hose to the drain valve at the bottom of the heater and run the end of the hose to a floor drain or out a window.**
3. **Turn on the drain valve and open the hot water faucet at the bathtub or laundry tub.**

To refill the heater after servicing, follow these steps:

1. **Turn off the drain valve and the tub or laundry faucet.**
2. **Go to the hot water faucet that is farthest from the heater and open this faucet to allow the air to escape from the system.**
3. **Turn on the cold water supply valve on the top of the water heater.**

   When water runs from the faucet you turned in Step 2, the heater is full of water. Go around to the other faucets and open them to allow any trapped air to escape. You're back in business.

Part V

The Part of Tens

In this part . . .

This part contains thirty bits of useful information that we think you want to know: how to maintain a septic system, how to conserve water, and where to find great plumbing information on the Web.

Chapter 19

Ten Fascinating Facts about a Septic System

In This Chapter

- Understanding your septic system
- Keeping your system running smoothly

Septic systems are private sewage disposal systems that are common in areas where a public sewage system isn't available. When properly designed and constructed, a septic system can last for 30 years or longer.

Septic System Parts

Septic systems operate naturally without the aid of pumps or any mechanical devices. A septic system consists of the three following parts, all of which are buried underground and each with an important function.

- **Septic tank:** The *septic tank* is an underground vault made from concrete, plastic, fiberglass, or steel. Its job is to hold and store solid waste while allowing liquid waste to flow into the system, where it leaches into the surrounding soil. The septic tank is where bacterial action breaks down the solid waste into liquid and sludge. The liquid flows into the system and the sludge settles to the bottom of the tank.
- **Distribution box:** The *distribution box* is a connection point for a series of pipes that carry the liquid waste to the leaching field — not unlike a railroad station or switching yard. Liquid waste flows into the distribution box and then flows through pipes into the leaching field.
- **Leaching field:** The *leaching field* is where liquid waste flows into perforated pipes and then disperses down into the surrounding soil. Local building codes have strict guidelines concerning the length of leach lines in septic systems. These guidelines are based on the type of soil in the surrounding area (rocky, clay, sand, and so on), depth of ground water, and the location of other water sources (streams, rivers, and lakes).

Pump the Sludge

Most systems require pumping of the sludge every three or four years. (This must be done by a licensed professional — look in your phone book under "Septic Systems, Pumping.") If your septic system requires more frequent pumping, consider cutting back on water usage by installing low-flush toilets and having the distribution box and leach lines cleaned.

No Kitchen Garbage

Kitchen garbage disposal units — a popular accessory with many kitchen sinks — shouldn't be used when you have a septic system. A septic system is for waste water and solids, not garbage. Kitchen waste interferes with the normal bacterial action of the system. For the same reason, never dispose of chemicals, grease, petroleum products, or anything else that could disrupt the natural bacterial action of your septic system.

Homeowners with a septic system who insist on having a disposal unit should have a septic system cleaning service pump their tank yearly. For larger families, this is especially important.

And don't forget about Mother Nature — composting is a natural alternative to having a disposal unit.

See No Evil, Smell No Evil

As a rule, if you smell sewage or see sewage rising above ground, you're faced with serious problems in your septic system. A good septic system is one that you don't even know is there.

Leave the Field Alone

Septic systems can fail if any part of the system is damaged by driving vehicles over the leach field. Tree roots can also clog a system and stop the flow of liquid waste — a professional drain-cleaning service should be able to correct this problem.

Chapter 20

Ten Easy Ways to Save Water

In This Chapter

- Saving water in the house
- Cutting down on water use outside the house

Here are ten (or so) easy-to-live-with changes to your water-guzzling habits. Nothing too dramatic; just sound advice on ways to use water more judiciously.

Take Showers

If you normally take a bath every day, substitute a shower a few times a week. When showering, shorten the time that you let hot water run down the drain. Use a shower restrictor, which restricts the flow of water, on a showerhead (under $20), and you can cut conventional showerhead water use by one third. We first used a restrictor on our sailboat, which has a small hot water tank. It works like a charm on board and at home.

Don't Leave Faucets Running

When you use a faucet, don't leave the water running. Turn it off while you brush your teeth, shave, or wash and rinse dishes.

Load Dishwashers Efficiently

Load dishes in dishwashers properly and carefully so that the unit works to its optimum. Also use the proper cycle; for example, don't use the heavy-duty cycle if the dishes only require a quick wash. Don't prerinse dishes if they don't need it. When buying a new dishwasher, choose one that has low water usage —in the seven to nine gallon range — because 80 percent of energy consumed by a dishwasher is for heating water.

Wash Only Full, Cold-Water Laundry Loads

In your clothes washer, choose low temperature or cold-water washing cycles, and wash full loads whenever you can. For smaller loads, adjust the setting for lower water.

Fix Leaky Faucets

Fix leaky faucets promptly to avoid expensive wasted hot water (see Chapter 10). And add a faucet aerator to faucets (see Chapter 15) to reduce the amount of water that flows from them.

Insulate the Hot Water Heater and Pipes

Insulate the hot water tank and pipes using inexpensive and easy to install blankets and wraps designed for them. And drain sediment from hot water tank every three months so it works efficiently (see Chapter 18).

Water the Outdoors Judiciously

When watering your lawn and garden, immediately repair leaks in the hose. When using sprinklers, adjust them so that they water only the ground, not the sidewalk and street or the house. Water in the early morning or evening — not during the heat of the day — so that the water soaks into the ground instead of evaporating into the air.

If you have an automatic system, maintain it by cleaning the heads and making sure stones or plantings don't obstruct them. Use timers to make sure that you don't overwater and waste water. And don't forget the basics: Good, old-fashioned weed pulling, on a regular basis, is still one of the best ways to maintain your garden and lawn.

Wash the Car with a Bucket

When you're washing the car, use a bucket of water or a hose with an automatic shut-off nozzle, instead of letting the hose run.

Chapter 21

Ten Plumbing-Minded Web Sites

In This Chapter

- Finding plumbing information on the Net
- Discovering our favorite Web sites

While you're surfing the Internet, you just may be in the mood for some plumbing power. We suggest these ten great plumbing Web sites.

Housenet

You're probably not surprised that we recommend `www.housenet.com`, because we created the site! We're mighty proud of one feature called Handy-at-Home Tips where you can find hundreds of good ideas and advice from visitors — real people, with real solutions.

Fluid Master

Meet Dr. Flush and find out how a toilet flushes at `www.fluidmaster.com`, a nifty site created by the manufacturers of toilet repair kits.

NAPHCC

Go right to the pros to find good advice about hiring a plumbing contractor at `www.naphcc.org`, the Web site of the National Association of Plumbing, Heating, and Cooling Contractors.

Plumbing Heating & Cooling Information Bureau

The Plumbing Heating Cooling Information Bureau at www.phcib.org is a good resource for basic plumbing check-up and maintenance information.

National Kitchen and Bath Association

The National Kitchen and Bath Association site at www.nkba.org features helpful planning advice for remodeling kitchens and bathrooms. You can find extensive information about design advice.

Lowe's

You may already shop at a Lowe's Home Improvement Center, but they have a nice Web site, too. Check out www.lowes.com for some practical information about plumbing problems that may arise — like frozen pipes or emergency repairs.

Dap

At the U-Fix-It House at www.dap.com you can find basic information about where and why to use their different types of caulk and sealant. Just click on the Do-it-yourselfers area and you find room-by-room how-to projects.

Kohler

When you want to do some wish-book shopping for plumbing fixtures, check out www.kohlerco.com. The people at Kohler do a nice job of featuring their products, and include check-off lists for them.

Moen

If you're planning to remodel a bathroom or kitchen visit www.moen.com, one of the largest manufacturers of faucets. They feature a nice selection of articles about decorating and remodeling, which you can use to get you started.

American Standard

Another manufacturer site worth visiting is American Standard at `www.us.amstd.com`. Here, you can find a nice array of their product line to help you plan a bathroom remodeling job.

Index

• A •

• B •

• C •

• D •

• E •

• F •

• G •

• H •

• I •

• L •

• M •

• N •

• O •

• P •

• Q •

• R •

• S •

• U •

• V •

• W •

IDG BOOKS WORLDWIDE BOOK REGISTRATION

We want to hear from you!

Visit **http://my2cents.dummies.com** to register this book and tell us how you liked it!

- Get entered in our monthly prize giveaway.
- Give us feedback about this book — tell us what you like best, what you like least, or maybe what you'd like to ask the author and us to change!
- Let us know any other *...For Dummies*® topics that interest you.

Your feedback helps us determine what books to publish, tells us what coverage to add as we revise our books, and lets us know whether we're meeting your needs as a *...For Dummies* reader. You're our most valuable resource, and what you have to say is important to us!

Not on the Web yet? It's easy to get started with *Dummies 101*®*: The Internet For Windows*® *98* or *The Internet For Dummies*®, 5th Edition, at local retailers everywhere.

Or let us know what you think by sending us a letter at the following address:

...For Dummies Book Registration
Dummies Press
7260 Shadeland Station, Suite 100
Indianapolis, IN 46256-3917
Fax 317-596-5498